Astronomy
for the
Utterly Confused

Terry Jay Jones, PhD
Jeanne K. Hanson

McGraw-Hill

New York   Chicago   San Francisco   Lisbon   London   Madrid
Mexico City   Milan   New Delhi   San Juan   Seoul
Singapore   Sydney   Toronto

520
JON

The **McGraw·Hill** Companies

Library of Congress Cataloging-in-Publication Data

Jones, Terry Jay.
    Astronomy for the utterly confused / Terry Jay Jones, Jeanne K. Hanson.
—1st ed.
       p.   cm.
    Includes index.
    ISBN-13: 978-0-07-147158-9
    ISBN-10: 0-07-147158-8
    1. Astronomy—Popular works.   I. Hanson, Jeanne K.   II. Title.

QB44.3.J66   2007
520—dc22   200603247

1 2 3 4 5 6 7 8 9 0 DOC/DOC 0 1 2 1 0 9 8 7 6

ISBN-13: 978-0-07-147158-9
ISBN-10: 0-07-147158-8

This book was set in TimesTen by International Typesetting and Composition.

Printed and bound by RR Donnelley.

McGraw-Hill books are available at special quantity discounts to use as premiums and sales promotions, or for use in corporate training programs. For more information, please write to the Director of Special Sales, McGraw-Hill Professional, Two Penn Plaza, New York, NY 10121-2298. Or contact your local bookstore.

This book is printed on acid-free paper.

This book is dedicated to Jane Rauch Kitchen, astronomy major at Wellesley College, 1932–1936.

Contents

Part IV   Stars and Galaxies

Preface

The purpose of *Astronomy for the Utterly Confused* is to enhance readers' interest in and knowledge of basic astronomy. The book ranges from the most energetic visible objects in the universe, quasars, through the planets of our solar system, to the infinitesimal sub-atomic nuclei of hydrogen and helium which power the nuclear fusion of our Sun. It moves from the Big Bang through the lifetimes of galaxies to the end of the universe, as presently conjectured. Mentioned along the way are some of the most active major research issues in the field.

With its quick tips, boxes, schematic drawings, photographs, quizzes, and text, this book incorporates 18 chapters discussing the following subjects: the universe in time and space, the Earth's place, our vantage point on Earth, key concepts and basic laws, the solar nebula and the formation of gas planets, the solar nebula and the formation of rocky planets and moons, the smaller bodies of our solar system, the Sun's role in our solar system, origins of the puzzles of space-time and gravity, down to the atomic and sub-atomic levels, how these are studied, how stars evolve, how they end their lives, the formation and evolution of galaxies, the role of dark matter, the role of dark energy, expansion/contraction/dissolution of the universe, and space travel.

Acknowledgment

The authors would like to thank our editor, Barbara Gilson of McGraw-Hill. Many editors are knowledgeable, many are thorough, and many are a true pleasure to work with. Barbara stands out in all departments.

Organization of the Text

- Astronomy moves between the largest scales and the smallest. Though the book begins with large-scale phenomenon, then pauses to move to the sub-atomic level, then returns to the large scale, some readers may wish to begin with Chapters 9–11, where the atomic and sub-atomic scales are described.

- Those who are not reading the book all at once may wish to consult Chapter 4, "Key Concepts and Basic Laws," regularly, for both perspective and a refresher on details.

- Each chapter includes quick tips, boxed sidebars, and a three-part quiz.

PART I

Where Are We? The Basics

CHAPTER 1

The Universe in Time and Space

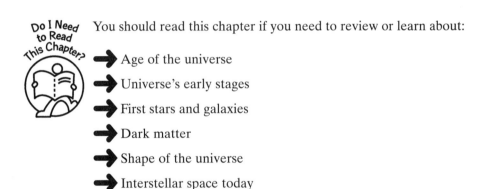

Do I Need to Read This Chapter?

You should read this chapter if you need to review or learn about:

➡ Age of the universe

➡ Universe's early stages

➡ First stars and galaxies

➡ Dark matter

➡ Shape of the universe

➡ Interstellar space today

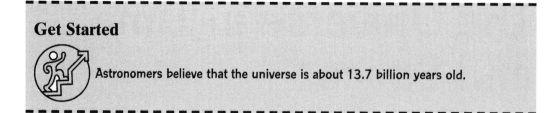

Get Started

Astronomers believe that the universe is about 13.7 billion years old.

Age of the Universe

The universe as we know it had a beginning. As readers might imagine, it can be a bit difficult to discern the date of this event. Astronomers analyze the universe's age in two main ways: by discovering the ages of the oldest stars and galaxies and by measuring the Hubble constant, which provides information about the rate at which the universe expands. Also used to solve this riddle are measurements of the cosmic microwave background—a kind of echo of the universe's birth—and of elements in interstellar gas. These results seem to be in good agreement: the universe is about 13.7 billion years old.

In just one of the research techniques used to decide upon the ages of the stars, astronomers have been able to mimic in laboratory environment some of the conditions inside stars which are aging now. As stars consume hydrogen and helium fuel at the intense heat and pressure of their nuclear fusion, they gradually burn out, no longer able to emit the same amount and the same types of radiation. Using a particle accelerator, which fires a proton at the nucleus of a nitrogen-14 atom, astronomers can estimate how long that particular aspect of nuclear fusion takes inside a star as it gradually runs out of gas. With the use of this method, it turns out that the very oldest stars are just under 14 billion years old. This result makes for an excellent fit with the age of the universe as determined in the other ways mentioned above, and it is also a good illustration of how large-scale phenomena, such as galaxies, can be studied by investigating small-scale phenomena, such as subatomic changes.

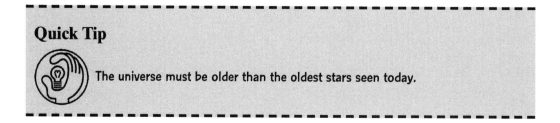

Quick Tip

The universe must be older than the oldest stars seen today.

Universe's Early Stages

The first 400,000 years in the life of the universe were extraordinarily violent. In its birth, called the Big Bang, mass-energy was created and began immediately to expand, racing out at an almost unimaginable speed. Do not picture a large firecracker lighting up space. There was no space, not until the Big Bang itself created it.

Immediately after the initial Big Bang—before even a few nanoseconds had passed and before matter had become separated from energy—the universe inflated. In this inflation period, it suddenly expanded at an even more rapid rate, ballooning from subatomic size to grapefruit size in about one trillionth of a second. This extreme expansion smoothed out any large clumpings, or variations in the universe's density, leaving only small fluctuations. This is why the universe today looks similar in every direction.

After the epoch of inflation, but still within the first few microseconds of the new universe, the earliest matter was created: quarks and gluons. These were in a fiery hot form of matter and energy called plasma, a unique state that resembles both a gas and a liquid. In it, matter and antimatter were continuously created and destroyed through high-energy processes. Soon, a small amount of matter became established, winning the battle, in a sense, over antimatter. That also set the stage for the universe we know today.

Next Stages

As the plasma expanded and cooled in all directions, some small random fluctuations remained within it. The simplest constituents of matter observable now formed from these fluctuations: the protons, neutrons, and other sub-atomic particles. Helium, a primordial gas, also began to form about during this time. This period of the early universe, called the era of nucleosynthesis, began about 0.001 seconds after the Big Bang and lasted until about 3 minutes after it.

Cosmic Background

Over the next 300,000 years, much less happened, relatively speaking. During this time, though, radiation was pushing out and gravity pulling in, creating a "tug of war" at cosmic scale. The young universe gradually cooled enough so that the new elementary particles, protons and electrons, became bound up into hydrogen atoms. By the time this process was complete, the universe was about 350,000 years old. Matter and light were now separate.

The "echo" of this light from the Big Bang can still be detected today, some 13.7 billion years after the universe's birth. The light is in a form of radiation, a kind of faint glow, called the cosmic microwave background. Orbiting space probes and balloon-borne instruments have mapped this dim radiation over the whole sky, detectable only as microwaves. With a temperature of just 2.73 degrees Kelvin, the universe's birth echo is only a couple of degrees above absolute zero.

This image of the sky was made from observations of the cosmic microwave background by the Wilkinson Anisotropy Probe (WMAP). The fluctuations in intensity can be used to determine many properties of the Universe, including its age. NASA

Quick Tip

1. Only in an immense universe created in an immense event would echoes of something that occurred 13.7 billion years ago still be detectable today.

2. To hear a CD that simulates the sounds of the Big Bang and the universe's first 100 million years, go to www.astro.virginia.edu/~dmw8f.

First Stars and Galaxies

As was mentioned in the previous section, there were still some irregularities in the density of the universe, which remained after the period of inflation. The universe was more like a miso soup than a completely clear broth. These differences, though, were minimal—one small area was never more than 0.01 percent brighter and hotter than another. But astronomers think that this slight difference, a kind of texture in the otherwise smooth distribution of matter, led to the first generation of stars and that this occurred only 400 million years after the Big Bang. However, the matter that stars and planets are made up of, normal matter, was too thinly dispersed to start gathering under the force of its own gravity. Some other source, a type of matter physicists now call "dark matter," was necessary. It is described later in this chapter and in Chapter 15, The Role of Dark Matter.

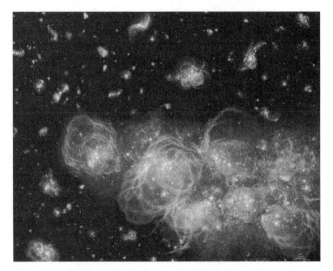

Artist's conception of the Universe when the first stars were formed. The large bubbles and arcs represent the aftermath of the first supernova explosions, blasting holes in the pristine interstellar medium. Courtesy NASA-Adolf Schaller

First Galaxies

After these earliest stars, the first galaxies were formed, which are organized aggregations of stars bound together gravitationally. The oldest galaxies and quasars discovered date very far back, to just 800 million years after the Big Bang. (Quasars are unusually bright, powerful objects, which get their power from immense black holes; they are discussed in Chapter 14, Galaxy Formation and Development.) Astronomers can tell by the composition of quasars that they are "second generation" at least; they contain material such as iron and magnesium that had to have been created by nuclear fusion inside an earlier generation of stars. Hence, they think that we will later detect even older such objects, even farther away in time and space, and that will push the occurrence of the first galaxies and quasars back even closer to the Big Bang. Some primitive galaxies, under study now, may already be expanding our understanding in this way.

Youthful Universe

The youthful universe, just 1 billion years of age, was a constantly active place. Gravity was steadily pulling in material to form larger, denser bodies and groupings of them all the time. This period of intense galactic formation continued in

The more distant a galaxy is, the longer it takes its light to reach us. We see very distant galaxies at a much earlier epoch in the Universe. These Hubble Space Telescope images show different galaxies at different times in the age of the Universe. NASA

full swing for a few billion more years. Galaxies also continue to change today, of course, and sometimes in dramatic ways (to be described in the next chapter and in Chapter 14, Galaxy Formation and Development) but completely new galaxies no longer seem to be forming today in our universe.

By about 1 billion years after the Big Bang, the first black holes had also formed. These objects are so dense, and with such power in their gravitational pull, that not even light can escape them. Their existence requires mature stars, galaxies, clouds of gas, and all the main visible elements of the universe today,

Quick Tip

The farther away an object is in space, the older it is. This is because light takes time to travel through space—moving at the rate of 186,411 miles (300,000 kilometers) per second. And the longer that light has taken to reach us, the farther away the object must be. One *light-year* is equal to a distance of about 6 trillion miles. (It is not a measure of time but of distance.)

since these are both their origin and their "food." (Black holes are discussed further in Chapter 13, The Deaths of Stars.)

At 1–2 billion years after the Big Bang, about half of the existing stars we see today had formed, swirled together with other material, and been knit into complex galaxies. Some of these galaxies, far from being primitive blob-like gatherings, showed spiral structures like the Milky Way does today. Some were also quite massive and halos of dark matter surrounded them by this stage already.

Within a few billion years, most of the galaxies in the universe had become established. The universe had become mature. Even now, however, it is far from static, as readers shall see throughout the book.

The Hubble Deep Field. The faint smudgy galaxies seen in this image are much more distant and much younger than our Milky Way galaxy. NASA

Dark Matter

This exotic substance—primitive matter that has never interacted with light and is invisible to us—is required to hold together all the galaxies and clusters of galaxies that we see populating the universe. Without it, the galaxies would never have been able to form in the first place. Dark matter is like invisible binding or hidden glue. (It is discussed further in Chapter 15, The Role of Dark Matter.)

Astronomers now believe that the universe is about 4 percent ordinary matter, 20 percent dark matter, and 76 percent dark energy. (Dark energy is discussed in Chapter 16, The Role of Dark Energy.) A galaxy like our Milky Way is merely a point of light within a cocoon of this dark matter and subtly inflected with a great deal of dark energy.

Astronomers have found evidence of dark matter in several ways. Its presence has been detected in the cosmic microwave background, the echo of the Big Bang. Its footprint has also been found amid clusters of galaxies, as it slows down the speeds of the galaxies orbiting around the center of a galactic cluster. Evidence of it is seen, too, in the x-rays emitted by the hot gas between galactic clusters. It can also be surmised from the rotation curves of galaxies and its pull can be seen in the way the gravity of a cluster of galaxies bends light. It simply must be there. But the nature of it is a major research puzzle in astronomy today.

The distant cluster of galaxies MS105. Large clusters such as this one are now known to be held together by dark matter, not the matter in stars. NASA

The long, stretched out arcs in this picture are the distorted images of background galaxies shining through the relatively nearby galaxy cluster Abel 1689. The images are distorted by gravitational lensing caused by the dark matter in the cluster. NASA

A typical cluster of galaxies contains a few dozen galaxies, bound together gravitationally, extending across about 1.4 million light-years from side to side.

Shape of the Universe

Is the universe finite or infinite in extent? Is it flat or curved? Does it have the shape of a saddle? Or a 12-sided object called a dodecahedron? Or a 3-torus (made by gluing opposite sides of a box together)? Or something else? And how on earth (or anywhere else) would astronomers be able to figure this out, since from here, on our tiny planet, the universe looks as though it goes out equally in all directions, like a sphere, which may extend forever? We certainly cannot see any edge to it, or beyond it.

One method astronomers use to try to figure this out is, again, by analyzing the cosmic microwave background. Within its blur still are faint waves, just as a pebble thrown in a lake creates waves that spread outward and yet remain visible. One of the waves detected is not as long as the rest, as though something were stopping it. Is that perhaps the edge of the universe? The puzzle of this wave and other approaches to the size and shape of the universe will be discussed in more detail as the book proceeds.

Interstellar Space Today

Besides the more obvious stars—about 100 billion of them or more in each galaxy in the observable universe—and the galaxies, which also number at least 100 billion discovered so far, visible space is cluttered with clouds of gas and dust. Much of this is the effluent of dying stars. It also becomes the raw material for new ones. Within these interstellar gas clouds, created by the chemical reactions within them, are hundreds of complex, organic molecules. They are organic because they contain carbon, which means that they are the building blocks of life as we know it. Space is not empty. Other gas and dust clouds, found usually in the peripheries of galaxies, are more ancient, left over from the masses of primordial hydrogen and helium formed in the first eras after the Big Bang.

The organic material from space reaches us, and in a way most people would not expect: it falls down to Earth with the meteorites, comet debris, and general cosmic dust that sift down to the ground level of our planet almost constantly. Anyone who stays outside the length of time a soccer game lasts, will go home with cosmic dust in their hair.

Our planet Earth formed about 4.6 billion years ago, and by about 3.5 billion years ago, it was already teeming with the life that may have begun as the organic molecules in interstellar space. This earthly life was primitive then; microbes, which though they did not breathe oxygen, were yet most definitely alive. Our planet was not clothed in an oxygen atmosphere until about 2.5 billion years ago. Did life come from the stars? The answer is almost certainly yes. (And no little green people were needed to bring it.)

Everything came ultimately from the stars: the chemical elements we encounter on our planet, from carbon to iron, nitrogen to oxygen, were created in these fiery infernos. This is recycling on the cosmic scale. Before the stars, only the simplest elements, hydrogen, helium, and a bit of lithium existed. All the other elements have been cooked up in their "furnaces" and exploded out into space when they changed or died.

It's a Wrap   In this chapter you have learned that:

✔ The universe is about 13.7 billion years old.

✔ The inflationary period began even before the universe was a few nanoseconds old.

The first substance in the universe, a quark-gluon plasma, formed in the first few microseconds.

 The first stars had formed by the time the universe was about 200 million years old.

 The first galaxies and quasars had formed by the time the universe was 800 million years old, perhaps even sooner.

Test Yourself

True/False Questions

1. The universe expanded out uniformly in all directions in its inflationary period.
2. The echo of the Big Bang is still immensely hot.
3. The first generation of stars formed before the galaxies did.
4. Black holes formed before the galaxies did.
5. There is more dark matter than matter that can be seen.

Completion Questions

1. A plasma is the _____ substance known to exist.
2. The youngest star detected so far in today's universe has a mass about _____ the mass of the Sun.
3. Black holes had formed by about _____ years after the Big Bang.
4. The universe is about _____ percent dark energy.
5. Before there were stars, the only elements that existed in the universe were basically _____, _____, and _____.

Multiple Choice Questions

1. The cosmic microwave background is what kind of a feature of the universe?
   a. One which makes microwaves work on Earth
   b. The force at the center of the universe
   c. An echo of the Big Bang

2. The density differences between areas in the original plasma, which led to the formation of all the stars and galaxies, were about how large?
   a. 10 percent
   b. 0.01 percent
   c. 1 percent

3. How do we encounter the organic molecules from interstellar clouds?
   a. They have been found on the Moon by the astronauts.
   b. They fall to Earth on meteorites and in comet and other cosmic debris.
   c. They are now used to make organic food.

4. Dark matter is needed to do what?
   a. Hold the universe as we know it together
   b. Create dark energy
   c. Blot out some of the stars in the Milky Way
5. The Earth formed about how long ago?
   a. 3.6 billion years ago
   b. 4000 years ago
   c. 4.6 billion years ago

ANSWER KEY

True/False
1. F   2. F   3. T   4. F   5. T

Completion Questions
1. hottest   2. 1/1000th   3. 1 billion   4. 73 percent   5. hydrogen, helium and a trace of lithium

Multiple Choice Questions
1. c   2. b   3. b   4. a   5. c

▶◀▶◀▶◀▶◀▶◀▶◀▶◀▶◀▶◀▶◀▶◀▶◀▶◀▶◀▶◀▶◀

CHAPTER 2

The Earth's Place

▶◀▶◀▶◀▶◀▶◀▶◀▶◀▶◀▶◀▶◀▶◀▶◀▶◀▶◀▶◀▶◀

Do I Need to Read This Chapter?

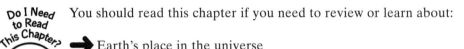

You should read this chapter if you need to review or learn about:

➡ Earth's place in the universe

➡ Earth's place in the solar system

➡ Earth as a moving body

Our Place in the Universe

Within the universe's billions of galaxies, most galaxies are part of a cluster with their neighboring galaxies, held together across the vast distances by gravitational attraction. These clusters are also part of superclusters, which are even larger aggregations of galaxy clusters. In between the superclusters, space is emptier.

A distant galaxy cluster around the giant elliptical galaxy 3C 324. Much of the galaxies in the universe are in large clusters like this one. NASA

The Earth is within a cluster and a supercluster called the Local Group and the Local Supercluster, respectively. Not exactly as local as the little neighborhood deli, our Local Group alone—about 40 galaxies strong, including the Milky Way—is about 7 million light-years across. Our Local Supercluster, composed of about 4000 galaxies, is about 60 million light-years across. (More about galactic groupings is found in Chapter 14, Galaxy Formation and Development.)

Almost all of the galaxies in this image are in the Virgo Cluster, the largest cluster of galaxies near to our location in the universe. We are on the periphery of the Local Supercluster, a very large cluster of galaxies centered on the Virgo Cluster. NASA

Astronomers believe that the universe may contain as many as 10 million superclusters, immense aggregations of matter which contain about 1000 galaxies or more apiece. Many of these are engaged in universe-scale collisions at any given time. Each of these meldings takes several hundred million years to become a merger. Shock waves from the events have been detected, even at 800 million light-years away. One is occurring at that distance from us, right now.

At a less immense scale, though still large almost beyond imagination, is our own galactic home: the Milky Way galaxy. Within this swirling spiral, the

An X-Ray image of the center of the Virgo Cluster taken by the Chandra X-Ray Observatory. The bright spot in the center is the giant elliptical galaxy M87. Most galaxy clusters have one or more giant ellipticals at their centers. NASA

An ultraviolet image of M33, a nearby small spiral galaxy that is a member of the local group along with the Milky Way, Andromeda and many smaller galaxies. NASA

Earth lies about 28,000 light-years out from the center, on one of the galaxy's outlying spiral arms, though but not especially near to the actual visible edge.

Movements

Within our Local Group, some of the galaxies are moving toward our Milky Way, some away from us, and some at angles. These complex motions come primarily because denser bodies act strongly by gravity on the less dense ones, and there are variations of density within galaxies and between galaxies themselves. So the gravitational forces act sometimes in concert, sometimes at cross purposes, and usually at intermediate angles.

Four of the smaller galaxies in our Local Group are engaged in a more forthright attraction. They not only orbit the Milky Way but seem to be in the process of being consumed by us, their gas clouds pulled in by the Milky Way like spider webbing reeled in and their stars captured. The names of these four "crime-victim" galaxies are the Large Magellanic Cloud, the Small Magellanic Cloud, Canis Major, and Sagittarius.

Even within our own smaller neighborhood of the Milky Way, individual stars are moving in many directions, drawn by the gravity of denser, nearby stars and by other forces that affect motion. But, at the same time, most stars are securely orbiting around the center of the galaxy itself, under the sway of its immense mass. The Sun's orbit (along with the rest of our solar system) around this galactic center requires 230 million years for one trip.

An all sky image from the 2MASS survey showing the plane of our Milky Way galaxy. The bright area in the middle is the Galactic Center, about 7500 parsecs away. The two bright patches to the lower right are the Large and Small Magellanic Clouds, the nearest sizeable galaxies to us. NASA-2MASS

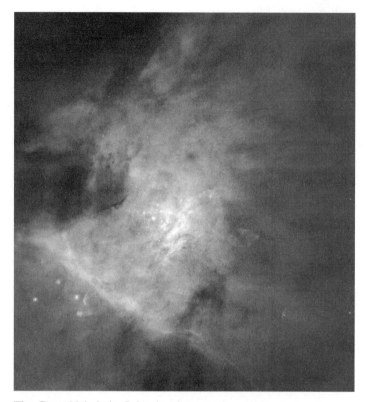

The Great Nebula in Orion is a large region or gas, dust and newly forming stars not too distant from the Sun, about 500 parsecs away. In perspective, the Galactic Center is about 15 times further away. NASA

These movements are all extremely complex and not at all completely understood. A star could certainly happen by our solar system unexpectedly, and there are runaway stars (as is described in Chapter 12, Star Formation, Star Development). There are even renegade black holes (see Chapter 13, The Deaths of Stars). Also, our path around the galactic center could easily bring us nearer to some unusual situation, since inner zones orbit the galactic center faster than outer zones, shuffling the "neighborhoods." But the distances are so vast that effects on us would almost certainly be nonexistent or negligible.

To add even more complexity to the movements, it is also true that all the galaxies that are outside of our Local Group appear to be moving away from us. And the farther a galaxy is, the faster it seems to be moving away. The same phenomenon would also be noted by observers, if any, in all other galaxies. The reason for this is discussed at the end of this chapter and its implications in Chapter 17, The Future of the Universe.

Our Place in the Solar System

Down at the more easily conceivable scale of our own solar system, the Earth is the third planet out from the Sun, farther from our star than Mercury and Venus, and closer to the Sun than Mars, Jupiter, Saturn, Uranus, Neptune, and Pluto. (For the controversy over Pluto's status as a planet, see Chapter 7, The Smaller Bodies of Our Solar System.)

The Sun is a dynamic force in the life of Earth, and not only because its heat and light make life as we know it possible. Its solar output, or radiative strength, is not perfectly constant, varying a few percent, since the Earth formed 4.6 billion years ago.

Even when the Sun's luminosity is stable, it emits massive blasts of particles that bathe our planet constantly, called the solar wind. The Sun also experiences massive magnetic storms, and at those times the solar wind increases vastly in power. In January, 2005, for example, a solar flare erupted, blasting energetic subatomic particles at us more violently than any such eruption in 50 years. The amount of material arriving at the Earth from the Sun even in a more ordinary period of solar wind can be billions of tons of star stuff, arriving over several hours up to several days. When this material hits the Earth in large amounts, our atmosphere, satellite, and radio transmissions can become disrupted; the Northern Lights, the Southern Lights, and the aurora are created—a night circus in the sky for those of us living at higher latitudes (as close or closer to the poles than to the equator). For more on the Sun, including its effect on Earth, see Chapter 8, The Sun's Role in Our Solar System.

Earth, a Planet in Motion

As most people realize, the Earth spins on its axis, which is what brings us a day and a night every 24 hours. Few, though, realize that this rotation is at a speed of about 992 miles (1600 kilometers) per hour at the equator, and a bit less at higher latitudes. Most of us also know that the Earth orbits the Sun in about 365 plus days, though few realize the speed of this motion either. It is 66,960 miles (108,000 kilometers) per hour, on average, along our elliptical orbit.

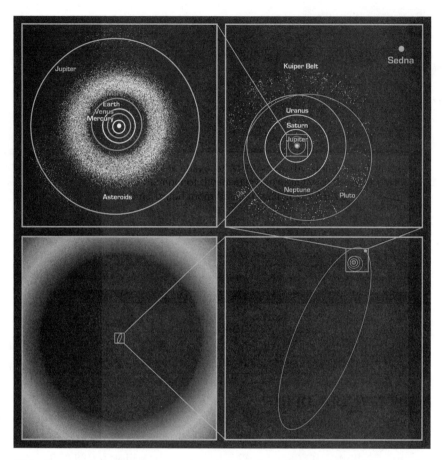

This shows a schematic representation of the region of space occupied by our solar system. The Oort Belt, the most distant area associated with our Sun, containing billions of comets is shown in the lower left. Each subsequent panel shows objects found in progressively closer to the Sun and the Earth. NASA

Our planet's other motions are also impressive. The Sun, taking our solar system with it, is also moving relative to neighboring stars, at about 43,400 miles (70,000 kilometers) per hour. We are also orbiting the center of the Milky Way galaxy at the speed of about 496,000 miles (800,000 kilometers) per hour. And, within our Local Group, the Milky Way is moving toward the Andromeda galaxy at about 186,400 miles (300,000 kilometers) per hour. In about 10 billion years, we could be very close. At that time, it is unlikely that any star will slam into any other star, though, since spaces between them are so vast. But comets, asteroids, and perhaps planets could have their orbits disrupted.

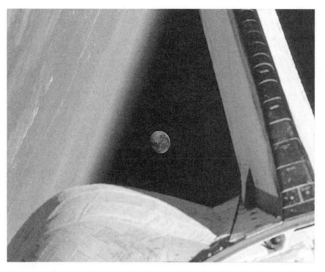

Humans have only recently been able to leave the immediate confines of our planet Earth. Here we see the Earth's Moon and the limb of the Earth as seen from the Space Shuttle. NASA

The Earth from space. NASA

Quick Tip

The reason that all galaxies appear to be receding from us is because the universe is expanding. Whether this will continue is the subject of Chapter 17, The Future of the Universe.

It's a Wrap

In this chapter you have learned that:

✔ The Earth is part of a solar system, a galaxy, a Local Group of galaxies, a Local Supercluster of groups of galaxies, and, of course, the universe itself.

✔ Galaxies interact with each other visibly through gravitational attraction, which includes collisions.

✔ Our planet, not standing still, is not only spinning on its axis and orbiting the Sun but also rushing through local space with the Sun, around the galaxy's center, and with the galaxy toward the Andromeda galaxy.

Test Yourself

True/False Questions

1. The Earth's position in the Milky Way is about halfway out from the center to the visible edge.
2. The universe is currently expanding.

Completion Questions

1. The motions between stars and between galaxies are primarily governed by
   _____.

2. The number of collisions occurring among the most massive superclusters of galaxies is about _____ at any one time.

3. Moving outward from the Sun, which planet is the next farther away from the Earth? _____.

Multiple-Choice Questions

1. Solar flares have what kinds of effect on Earth?
   a. They cause the ocean's tides.
   b. They make lightning.
   c. They can disrupt satellite and radio transmissions.

2. How long does it take the Earth to orbit the center of our Milky Way galaxy?
   a. About 10 billion years
   b. About 230 million years
   c. About 5000 years

ANSWER KEY

True/False
1. F   2. T

Completion Questions
1. gravity   2. 25   3. Mars

Multiple Choice Questions
1. c   2. b

CHAPTER 3

Our Vantage Point on Earth

Do I Need to Read This Chapter?

You should read this chapter if you need to review or learn about:

➡ The constellations and why they seem to change throughout the night and the year

➡ The planets and why they move differently from the stars

➡ The seasons of the year and how they happen

➡ Eclipses

Get Started

The constellations appear to move in a circle around the North Star every night and to change throughout the year, but this is only because the Earth itself is moving.

Constellations' Change Throughout the Night and Year

The constellations seem to change throughout the night. Anyone who has looked up twice or more at the sky in a single evening realizes that the constellations appear to have moved in a circle around the North Star from the point of view of those of us in the Northern Hemisphere. (This star is near to the Big Dipper, though closer to the zenith, or top, of the sky.) Each constellation seems to rise in the east and set in the west, making a large circle, though the ones closer to the horizon dip below it as part of their circle. The reason for these apparent motions is that the Earth is rotating from west to east, bringing a different starry scene into view as we twirl through the night. The motions seem even more complex because we are only seeing half of the night sky visible from Earth, the other half is visible to those of us in the Southern Hemisphere.

It simply could not be the constellations themselves moving as a group in this way. The reason is not only because the individual stars are so far away that we could never see such movements from here but also because these convenient groupings of stars—numbering 88 named constellations in the standardization of the International Astronomical Union (and with a different number and unique names in every culture)—are only artifacts. Most constellations are actually composed of stars that are actually nowhere near each other; astronomers call them asterisms. The stars are simply linked in our line of sight from the Earth, with some closer (perhaps only 10–100 light-years away) and others much farther away (up to hundreds of light-years distant). They do not move together at all.

The constellations appear to change throughout the year, too. Orion is not visible on summer evenings in the Northern Hemisphere, for example, The reason for this is that our planet is orbiting the Sun throughout the year, swinging us around to face different areas of sky.

Observers far from artificial light on Moonless nights, and who allow their eyes to become accustomed to the darkness can see up to 3000 stars on a clear night.

The constellation of Ursa Major, the Great Bear. The lines connect the stars we see in the sky to help illustrate the constellation. Notice the Big Dipper, not a constellation itself, is part of Ursa Major. TJJ

Quick Tip

Constellations composed of stars that are not actually near each other in the galaxy are called asterisms. They only appear together in our line of sight.

Especially in summertime, part of the Milky Way galaxy itself can be seen. It looks like spilled milk arcing across the zenith of the sky from a Northern Hemisphere viewpoint and a large smudge to those in the Southern Hemisphere. A handful of other objects observers can see with the naked eye are not stars at all, but rather our neighboring planets and an outer galaxy or two, depending upon the time of year and one's location. Andromeda galaxy can be seen from the Northern Hemisphere as a delicate blur, though it is 2.5 million light-years away. The Large and the Small Magellanic Clouds, both small, local galaxies, are visible from the Southern Hemisphere. Except these galaxies, all of the sky objects visible to the naked eye from Earth are in our Milky Way galaxy. And what we are seeing is only a tiny percentage of some 100 billion stars in our galaxy. (Various kinds of telescopes can, however, see much more. Chapter 18, Frontiers in Astronomy, describes some of the telescopes used.)

People who live in a city of the size of New York, however, can see no stars at all. Venus, Mars, Jupiter, and the Moon, are all that are potentially visible. There is too much artificial light.

Quick Tip

One way that astronomers identify the "addresses" of sky objects is by degrees, arc minutes, and arc seconds, conveniently using the dome-like appearance of the night sky and angles between it and the horizon. An example: 35°27′15″ means 35 degrees, 27 arc minutes, and 15 arc seconds—a position about a third of the way up between the horizon and the zenith.

The Planets and How They Move

The planets visible to naked-eye observers on Earth—Mercury, Venus, Mars, Jupiter, and Saturn—appear to move great distances across the sky. This is because they are much nearer to us than the stars are, enabling us to see their motions.

These planets also sometimes seem to move not only forward but backward (retrograde motion) against the background of the stars. They are not actually backing up. Each planet orbits the Sun at a different speed and each of their orbits comprises either a smaller or larger path than ours is. So, being out of sync with us, they seem to catch up with us or fall behind.

The Changing Seasons of the Year

This pleasing aspect of life on Earth is not caused by anything in the sky itself but rather by the movement of the Earth. As we orbit the Sun, our axis (like a stick stuck in the North Pole and emerging out of the South Pole) is always slightly tilted. This tilt affects us a little like the tilt of a car taking a curve at high speed—it causes the Earth to lean a bit. When the Northern hemisphere is tilting away from the Sun we experience winter, and when it is tilted toward the Sun we have summer. (The Southern hemisphere is always tilted opposite, making the seasons there also opposite.)

This tilt of the Earth's axis is small, even at relatively high latitudes on our planet, but it does change the angle at which the Sun hits us. It does not take much of a change, in the degree of solar radiation received, to transform conditions from a 90 degrees Fahrenheit summer afternoon to a 30 degrees Fahrenheit brittle winter dawn. There is much less of this effect near to and at the equator, which is why throughout the year there are no huge differences in seasonal temperatures.

Unlike what some people think, the Earth's distance from the Sun is not what creates the seasons. We are actually slightly closer to the Sun in January than we are in July.

Eclipses

A total eclipse of the Sun happens when a full Moon moves between the Earth and the Sun at just the right angle to allow the Moon's shadow to fall onto the Earth. This shadow creates a short period of darkness and then moves on.

The shadow of the Moon on the Earth during a Solar Eclipse. NASA

A solar eclipse happens because of a major coincidence. The coincidence is that the angular distance—the proportion of the sky occupied by a body as we see it—is nearly the same for both the Sun and the Moon, from our vantage point on Earth. We have a large Moon which is quite close and so takes up a lot of sky. We also have a far more immense Sun, but it is very far away. That makes it seem to be no bigger than the Moon.

A total eclipse of the Moon, a lunar eclipse, is quite different. It occurs when the Moon moves entirely into the shadow cast out into space by the Earth. We cast this perfect shadow when the Sun, Earth, and Moon are lined up perfectly, with the full Moon on the other side of the Earth from the Sun.

Lunar Eclipse. The shadow of the Earth is partly covering the Moon. NASA

These eclipses would occur much more often if the Moon's orbit were positioned directly around the Earth's equator. Instead, the Moon orbits us as a slight angle to the equator and so often misses the shadowing.

In this chapter you have learned that:

✔ The constellations do not really swing across the sky every night or move around throughout the year; these apparent movements are actually caused by the Earth's own rotation and orbital movement.

Our human civilization creates a remarkable amount of artificial light, making it harder for us to see the sky at night from urban areas. This image is a composite made up of nighttime images of the Earth taken from space. NASA

✔ The Earth experiences seasons because the tilt of our axis allows more, or less, solar radiation to hit a hemisphere directly.

✔ Eclipses are caused by alignments among the Sun, Earth, and Moon and the fact that the Moon appears to be about the same size as the Sun to observers on the Earth.

Test Yourself

True/False Questions

1. There are more than 100 conzstellations listed officially by astronomers.
2. Some planets appear to move both forward and backward across the sky.
3. The Earth's axis holds the same position throughout our orbital year.

Completion Questions

1. A total solar eclipse is caused by the Moon's _____.

2. The Earth's orbit brings us closer to the Sun in _____ than in _____.

3. Constellations composed of stars that are nowhere near each other in the sky are called _____.

Multiple Choice Questions

1. The constellations appear to move each night (in the Northern Hemisphere) around what object?
   a. Big Dipper
   b. North Star
   c. Moon

2. How many stars can be seen by naked-eye observers on Earth under the most favorable conditions?
   a. 1 billion
   b. 2 1/2 million
   c. About 3000

3. The Sun and Moon appear to be about the same size in our sky because
   a. They are our two most important heavenly bodies.
   b. Of a coincidence.
   c. They cause eclipses.

ANSWER KEY

True/False
1. F   2. T   3. F

Completion Questions:
1. shadow   2. January/July   3. asterisms

Multiple Choice Questions:
1. b   2. c   3. b

CHAPTER 4

Key Concepts and Basic Laws

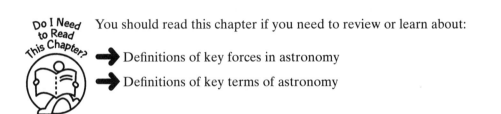

You should read this chapter if you need to review or learn about:

→ Definitions of key forces in astronomy

→ Definitions of key terms of astronomy

Definitions of Key Forces

If the universe were a play, the following would be the key characters:

- The interaction between matter and energy. This pas de deux began in the Big Bang and now governs everything from the smallest subatomic particle to the most powerful quasar or densest black hole.

  Matter can exist as a solid, liquid, gas, or plasma, and it can move between these phases, in changes called phase transitions. Matter exists as plasma at extremely high temperatures, as, for example, in the Big Bang and in the bodies of the stars, and plasma can also exist at cooler temperatures.

- Energy makes matter move, change, become transformed. And energy itself changes, from potential energy to kinetic or moving energy, to radiative energy such as the array of radiation from stars.

- Matter is made of atoms arranged in different ways, into different elements. The parts of an atom include *protons*, *neutrons*, and *electrons*. Within the protons and neutrons, found in the nucleus or center of each atom, are *quarks*. Outside the nucleus but within the atom are the electrons, forming a kind of cloud. Laws of physics that describe their behavior under different conditions are called the *weak force* and the *strong force*.

- The forms of electromagnetic energy emitted by astronomical bodies include gamma rays, x-rays, ultraviolet light, visible light, infrared light, and radio waves, depending upon their energy (with gamma rays the most energetic). Only the visible light can be seen by our eyes. But astronomers can detect the whole spectrum of radiation in the emission of a star, for example, by using different kinds of instruments.

- Scientists are looking for what they call the "Theory of Everything." In it, the interactions among the weak force, strong force, electromagnetic force, and gravity will be mathematically described.

- Gravity governs the large-scale motions of bodies in the universe. It is what keeps the Moon orbiting the Earth and is among the forces that bind whole superclusters of galaxies together. Surprisingly, it is the force of nature that is both the weakest and the least understood.

Definitions of Key Terms

Accelerating universe—the theory that posits that the universe will expand outwardly in all directions and at an increasing rate; galaxies will become farther away from each other

Antimatter—particles with the same mass as regular matter but with the opposite quantum number (antimatter and matter particles annihilate each other when they meet)

Astronomical unit (AU)—a distance equal to the average distance between the Earth and the Sun (about 150 million kilometers or 93 million miles)

Big Bang—an event that created the universe

Big Crunch—a possible (though unlikely) end of the universe if and when gravity overwhelms acceleration/expansion and the universe collapses back into itself

Black hole—an object so dense that the force of gravity overwhelms all other forces; so nothing ever reemerges once it is pulled in beyond the event horizon

Brown dwarf—an object that may or may not orbit a star (and so is not a planet) and that does not burn hydrogen in nuclear fusion (and so is not a true star); size range is 10–80 times the mass of Jupiter

Corona—the thin, top layer of the Sun's atmosphere, emits x-rays

Cosmic rays—subatomic particles from astronomical bodies and phenomena which rush through interstellar space at nearly the speed of light

Electromagnetic spectrum—the full array of types of "light": radio waves, infrared light, visible light, ultraviolet light, x-rays, and gamma rays (our eyes can see only the visible light part of the spectrum)

Elements—the basic chemical building blocks of the universe

Exoplanet—a planet found outside of our solar system

Grand Unified Theory (GUT)—the theory of astrophysics that accounts for the interaction of three of the four fundamental forces in the universe: the strong force, the weak force (which hold atoms together), and the electromagnetic force; gravity has not yet been incorporated

Hubble's constant—the rate of expansion of the universe, relates to Hubble's law that the more distant a galaxy is from any point in the universe, the faster it appears to be moving away

Infrared light—on the electromagnetic spectrum, the wavelength between radio waves and light visible to us

Kelvin—the temperature scale used by scientists

Light-year—the distance that light travels in 1 year; about 6 trillion miles (9.7 trillion kilometers)

MACHOs—massive compact halo objects which may be the dark matter that surrounds galaxies; may be made of objects such as planets and brown dwarfs

Orbital velocity—a mathematical formula that uses a star's distance from the galactic center and its orbital speed around it to gauge the total galactic mass within that orbital path (can be used for orbital velocities of other objects)

Parsec (pc)—a distance of 3.26 light-years

Protons—subatomic particles with a positive electrical charge; composed of three quarks

Quark-gluon plasma—the hottest substance ever created, it was formed in the Big Bang

Quasar—a very luminous galactic nucleus of the brightest type

Solar wind—charged particles streaming out from the Sun throughout the solar system

Spacetime—the combination of three-dimensional space with one-dimensional time into a four-dimensional system

Spectral class, spectrum—the light profile of a star, or the way its light appears when spread out by a prism inside an instrument called a spectrograph. The spectral classes of stars are organized by the main color within the star's light, which corresponds to their temperature, from the hottest to the coolest: O (violet-white stars); B (blue-white stars); A (white stars); F (yellow-white stars); G (yellower-white, like the Sun); K (orange stars); and M (red stars)

Speed of light—about 186,000 miles (300,000 kilometers) per second

Star—an astronomical body of gas that glows because of nuclear fusion in its core (the term is also used for dying or dead stars which no longer conduct nuclear fusion)

Subatomic particles—particles inside an atom (protons, neutrons, electrons, and their smaller components)

WIMPs—weakly interacting massive particles which may be the dark matter that surrounds galaxies; made of subatomic particles

PART II

Solar Systems and How They Form

CHAPTER 5

The Solar Nebula and the Formation of Gas Planets

Get Started

Our solar system's planets include the large, outer, gas giant planets Jupiter, Saturn, Uranus, and Neptune (also called the Jovian planets) the subjects of this chapter. To be discussed in the next chapter are the smaller, rocky planets also called the terrestrial planets those nearer to the Sun: Mercury, Venus, Earth, and Mars. (The case of Pluto and the outer reaches of the solar system is discussed in Chapter 7, The Smaller Bodies of Our Solar System.) Exoplanets, planets beyond our solar system, number about 200, and most of them discovered so far are gas giants. For that reason, they are discussed in this chapter.

Solar Nebula

As time marched forward in the young universe, stars were igniting, burning their fuel, and later releasing large amounts of material back into space. By the time a few million years had passed, this star debris, along with leftover gases and dust, which had not yet formed a star at all, was drawn by gravity into large, cold clouds. Later, one of these interstellar clouds of gas, dust, and ice a sphere probably a few light-years across was to become our own solar system.

Art its impression of a new solar system in formation. The swirling disk of gas, dust and rocks orbiting the central star will provide the raw material for the formation of planets. NASA

As this interstellar cloud swirled slowly, gravity held its loose materials together. Then under a disruption of some kind, probably a shock wave from an exploding star in the area, a part of the cloud began to collapse. When it did so, gravitational forces increased and the cloud heated up, especially at its center. It spun faster, and as the particles within it collided increasingly, it flattened into a disk. Called a solar nebula now, it was ready to begin forming a star and planets.

> The early universe contained myriad solar nebulas, and still does so today. This makes it highly likely that many solar systems have formed in the universe and are continuing to do so today.

Sun as Center

By the time the solar nebula had condensed to about twice the diameter of our solar system's today, its center had become very dense, hot, and churning. This heat was probably more than 1600 degrees Kelvin. Hydrogen and helium gases, which made up about 98 percent of the nebula overall, do not condense at all in heat this intense. But, at the center, the gas became dense enough to ignite into a thermonuclear fireball. It became a star, the one that we now call the Sun. This dates about 5 billion years ago.

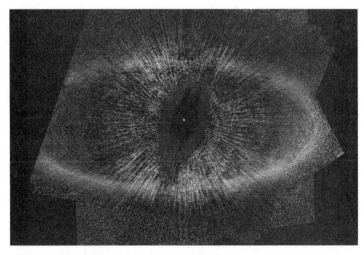

The star Fomalhault (HD 216956)has a ring of dust, rocks and debris orbiting around it left over from the formation of a solar system. This remarkable image was made by blocking the blinding light from the star with a chronograph on the Hubble Space Telescope in order to reveal the faint light from the ring. NASA

Earth

An image of the Sun in the light of the Hydrogen atom. The
Sun rests at the center of the Solar System and contains most
of the mass. NASA

Planets began to form also, in the cooler areas outside the nebula's center, some
closer to the new star and some farther away. The most accepted theory about
how planet formation occurred, called the *core accretion model*, posits that the
gases and rocky matter gradually built up more and more around core pebbles,
chunks, and then larger and larger chunks, colliding through gravity and then
being held together by it. The heat created in the collisions was enough to melt
the material so that it was able to meld together.

Not all astronomers agree with the core accretion model of planetary formation. A
significant minority believes that the gas giant planets formed much more quickly,
from a quick collapse of diffused parts of the nebula, perhaps in thousands or tens
of thousands of years instead of millions.

The new Sun was massive enough to cause the rest of the solar nebula to swirl
quickly around it. The inner areas became warmer as the rocky material condensed, while the lighter material such as the hydrogen and helium remained as
a gas. Gradually, even more of the rocky and metallic material melded in the

collisions that increasingly occurred. Here in the inner zones, the terrestrial planets, including the Earth, formed from these rocky masses. This occurred about 4.6 billion years ago. Farther out in the nebula, where it was cooler, gases and ice congregated into larger bodies. These then drew in a great deal of the remaining hydrogen and helium gas from the nebula. In this way the gas giant or Jovian planets formed.

Quick Tips

1. The newly formed Sun soon began to boil off the excess gases and dust grains that were present throughout the remnants of the solar nebula, engendering what is called its solar wind. (This wind still exists today, though its material is now entirely from the Sun itself.) The wind currents blew most of the remaining solar nebula material out into interstellar space. The solar system was now essentially complete.

2. Leftover bits of metal, rock, and ice in the middle areas of the solar nebula, around the gas giant planets, formed the asteroids and comets. (These smaller bodies are discussed in Chapter 7, The Smaller Bodies of Our Solar System.)

3. The moons of the Jovian planets Jupiter, Saturn, Uranus, and Neptune are mostly ice and rock, and thus more similar to the terrestrial planets than to the gas giants. So they are discussed with them in Chapter 6, The Solar Nebula and the Formation of the Rocky Planets and Moons.

4. The whole process of making a star and its array of planets seems to take about 10 million years from the beginning to the end.

5. The solar wind interacts with the Earth today in significant ways, from sometimes disrupting communications satellites to creating the Northern Lights and Southern Lights.

Solar System Motions

The Sun not only creates the light and heat throughout our solar system, which is both reflected and absorbed by all its planets and moons, but it also determines the orbits of the planets through its gravity. Most of the planets have elliptical though close-to-circular orbits around the Sun. They orbit it in the same direction and in the same plane, and they also rotate on their planetary axes in the same direction. These motions match what was surely the motion of the original solar nebula.

Exceptions are noted in later sections of the chapter and in Chapter 6, The Solar Nebula and the Formation of the Rocky Planets. (For more information on the roles of the Sun, see Chapter 8, The Sun's Role in Our Solar System.)

The Sun is in control because of its immense mass it contains more than 99.9 percent of the mass of the entire solar system. Our star burns its hydrogen fuel at temperatures of about 30 million degrees Fahrenheit. Made entirely of plasma, it is a powerhouse of nuclear fusion transforming about 600 million tons of hydrogen into about 596 million tons of helium, every second. The remainder is energy. This leftover is what makes life on Earth possible.

A montage of images of the nine traditional planets of the Solar System arranged in order from the Sun. Note that the relative sizes of the images are correct, but the distances between them are not. NASA

Quick Tip

1. Though the Sun is turning its own gaseous mass into energy, it is far from running out of fuel. Our star has been in technicolor action for about 5 billion years, and that is only about half of its lifespan. For how star lives end, see Chapter 13, The Deaths of Stars.

2. The gas giant planets have similar compositions and sizes and so are found together in the remainder of the chapter, considered one by one.

Jupiter's Main Features

The most massive planet in our solar system, Jupiter has storms in its atmosphere that are larger than the entire Earth. The most famous, a more or less permanent though shifting storm, is called the Great Red Spot. It is two or three times the size of our planet. Jupiter's mass is primarily dense, swirling gas, with only a small, solid core.

Almost a miniature solar system itself, Jupiter has at least 60 moons and a set of faint rings, which are also in orbit around it. As is described in Chapter 7, The Smaller Bodies of Our Solar System, it also controls its own set of asteroids the Trojan asteroids, and it exercises considerable gravitational influence upon the Sun's own asteroid belt lying between its orbit and that of Mars. The force of Jupiter's gravity regularly pulls a few of these asteroid belt objects off course and "kicks" them toward the inner solar system, sometimes right toward the Earth. Astronomers are monitoring this potentially hazardous phenomenon, as is described in Chapter 7, The Smaller Bodies of Our Solar System. (The website www.spaceweather.com also details this effort.)

Jupiter is the largest planet in the Solar System. If it had been about ten times more massive at birth, it would have become a low grade stellar companion to the Sun. As it is, Jupiter contains more mass than all of the other planets combined. NASA

A close up view of the famous giant red spot in the atmosphere of Jupiter. Notice the complicated swirls and flows of the gases in the upper atmosphere. NASA

An image of the polar region of Jupiter showing aurora, much like the aurora we see as the Northern Lights here on Earth. NASA

Quick Tips

1. Jupiter lies at an average distance from the Sun of 5.2 AU.

2. Its mass is more than 300 times that of the Earth, and its volume is more than 1000 times as large as ours.

3. Not solid, it is made up mostly of hydrogen and helium gases. Its small solid core does not begin until about 37,200 miles (60,000 kilometers) down.

Jupiter, a Failed Star?

Jupiter's composition is essentially identical to the Sun's; so it is possible to wonder why it has never ignited this fuel to become a star. The reason seems to be that it is not quite large enough. Astronomers estimate that if it were 80–90 times more massive, it would indeed ignite. Also remember that, it formed rather far away from the dramatically hot center of the solar system.

There are objects in the universe which are not quite stars but not quite planets. These are called brown dwarfs brown because they shine weakly and only in infrared wavelengths, and dwarf because they are smaller than stars. These are large objects one has been detected recently that is nearly five times as large as Jupiter and they range up to about the mass of 75 Jupiters. Jupiter is not large enough to be a brown dwarf.

The best way to characterize Jupiter is as a failed star and failed brown dwarf, but a very impressive planet.

Saturn, a World of Rings

The second most massive planet in the solar system, Saturn, is about 95 times the mass of the Earth. At an average distance from the Sun of 9.5 AU, it is about twice as far from the Sun as is Jupiter. Saturn is less dense than Jupiter, but it is also primarily made of hydrogen and helium gases around a small solid core. Its upper atmosphere is the site of fierce and quite variable winds.

Saturn's rings are the most spectacular of any in our solar system. There are seven major rings with more "ringlets" between them, occupying a zone vaster than that between the Earth and its Moon. Though these rings appear solid in a telescope, they are not. Rather, they are aggregations of many small particles and medium-sized bodies, all made of ice and rock, orbiting their planet relatively close together and reflecting enough light to blur into what appears to be solid from far away. The ring system is a flattened disk, just a few hundred meters thick. The origin was probably in a collision between Saturn and either a comet or asteroid, which smashed apart about 100 million years ago. All the pieces quickly fell into orbit around the giant planet, where they joined Saturn's 50-some moons (originating in similar ways). These moons act as "shepherds," their modest gravitational powers enough to hold the rings in place, and also as disturbers, sometimes creating ripples, waves, and gaps in the ring configuration.

The planet Saturn, with its wondrous system of rings. Notice the shadow of the rings on the sunlit side of the planet and the shadow of the planet itself on the rings on the night side. NASA

Quick Tips

1. Saturn has at least 50 moons, the main ones of which are discussed in the next chapter.

2. Depending upon how one counts, though, this planet could be considered to have billions more moons. If one were to add up all the pebbles, rocks, boulders, and barn-sized objects within its rings, they would total at least that.

Saturn, Cassini Spacecraft's Visit

Once the Cassini spacecraft from Earth entered Saturn's orbit in 2004, it began to transmit findings back. Detailed features have been seen in this planet's system that are as small as 23.56 miles (38 kilometers) across.

Among the Cassini discoveries were: an additional moon (small, at only 4.34 miles (7 kilometers) across), two more rings, major storms that create lightning a million times more energetic than any lightning on Earth, atmospheric storms that merge, and clouds of hydrogen and oxygen atoms made as Saturn's radiation breaks water apart. (The water is present because small bodies in the rings collided and released their ice.) Mysteries remain, however, exactly when the rings formed and how long they will last.

Saturn has Northern Lights, called aurora, which look the way the Earth's do. They may form like ours, when charged particles from the Sun's solar wind hit the planet's magnetic field, ionizing particles in its atmosphere. Or, astronomers think, the material energized here may instead be coming from material evaporating from one of Saturn's moons.

Like Jupiter, the upper atmosphere of Saturn shows bands of clouds and motions in the clouds, but not quite as strikingly as in Jupiter. NASA

Uranus as the Tipped-Over Planet

A pale blue-green gas giant planet Uranus is made of hydrogen, helium, and hydrogen compounds such as methane (the latter is bluish-green), partly frozen into an icy, gaseous sphere. Its average distance from the Sun is 19 AU and it is more than 14 times as massive as the Earth.

The most unusual feature of Uranus is that it is tipped onto its side, the north pole and south pole pointing out to its right and left instead of out its top and bottom. The cause of this extreme tilt was probably a massive collision with some other body as the solar system was forming long ago. Since then there has been no way for the planet to set itself right.

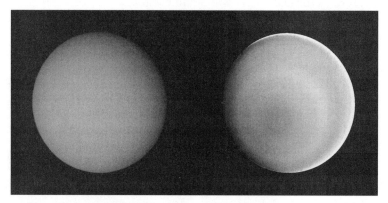

Uranus is a gas giant that spins with its axis of rotation lying almost on its side. This image compares a normal view of the planet with a contrast-enhanced view that shows circular motions around the pole. NASA

Quick Tip

Uranus has 13 faint rings and at least 27 moons.

Because of Uranus' unusual orientation, an observer seated at one of its poles would have 42 years of daylight followed by 42 years of darkness over its 84 year orbital year. (There are, however, no observers on this planet.)

Neptune, Slightly Smaller

Neptune, though 50 percent farther away from the Sun than Uranus, is similar to it in mass (about 17 times as massive as the Earth). It is, however, even bluer, a coloration that comes from the methane gas in the atmosphere.

Netpune does have seasons, perhaps surprisingly since it receives only 0.1 percent of the solar radiation that the Earth does. Each of its four faint seasons lasts about 40 years.

This gas giant has faint rings and at least eight moons. Its gaseous atmosphere features major storms and fierce winds.

Neptune is almost a twin of Uranus. Here we see some of the features in the upper atmosphere of the planet. NASA

Spacecrafts, all robotic, have flown by the planets often. But only on Mars and Venus, so far, have spacecrafts landed. And the Moon is the only body in the solar system to have received human as well as robotic landings.

Exoplanet Numbers and Variations

The planets that have been discovered beyond our own solar system are called exoplanets by astronomers. These now number about 200, with new and quite different other worlds being identified every year. No longer can anyone believe that we are the only solar system in the universe, though most of the systems found so far are not particularly like ours.

The exoplanets range widely, as can be seen even in a small sample of what has been identified. One of the youngest exoplanets seen so far is less than 1 million years old and about 420 light-years away, orbiting the star CoKu Tau 4. A nebula "nursery" of 300-some newborn stars has been found with what looks like disks of solar nebulas around two of them. The oldest exoplanet seen lies in a crowded globular cluster of stars some 12.5 billion years old, a setting in which astronomers did not think a planet could form or survive. The strangest exoplanet observed, about 200 light-years away, is a gas giant but possibly with a huge rocky core that weighs about 70 Earth masses. A whole exoasteroid belt has also been found, around a star much like ours. Yet another exoplanet, named HD149026b, is the densest one found; it may be 50–70 percent heavy elements as opposed to the more typical hydrogen and helium. A tiny star with one huge planet has been found. And another, OGLE-2005-BLG-390Lb is the one with the lowest mass detected so far, it weighs just 3–11 times what our Earth does. And, at just 63 light-years away, the planet orbiting the star Vulpecula has even had its emission of heat detected.

Several solar systems with three exoplanets apiece have also been detected: one system has been discovered with four planets, and, at only 42 light-years away, a system with three planets and an asteroid belt has been detected in orbit around a Sun-like star. A star-forming area in the constellation Orion, some 3,000 stars in a region only about 13 light years across is also especially interesting. The stars appear to vary in size from a few times as massive as Jupiter to about 40 times as massive as the Sun, shock waves are coursing through the region, and the potential for an array of exoplanet formation exists. More discoveries will probably be made as this book is being published.

Astronomers are able to find smaller and smaller candidate exoplanets all the time and are now occasionally able to detect those only about 10 times the size of the Earth.

Exoplanet Detection

One might imagine that it is hard to detect objects so much smaller than their stars and that do not even shine with their own light.

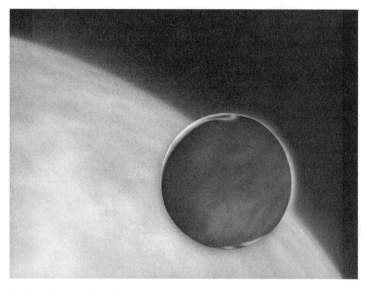

Artist's impression of a planet orbiting another star. NASA

Astronomers began this venture by looking at nearby stars to see if they "wobbled" a bit as they rotated or "wiggled," moving slight differently across the sky from our vantage point here. To make a star wobble or wiggle, a planet would not only have to be present but also quite large. Some exoplanets like these were indeed found and the race is on to find more and different ones. So far, though, we are finding mostly very large, usually gas giant, exoplanets because they are the ones we can best detect.

A second early technique was to look for a star being occulted its light blocked partly or completely by something in transit across the "front" of the star from our vantage point here. This might be an exoplanet passing in front of the star in our line of sight. These would also be, by definition, larger planets quite unlike Earth. The ones found this way do resemble Jupiter and Saturn far more than they do the terrestrial planets.

These techniques have now been refined considerably, with very small wobbles, wiggles, and occulations detectable. Wobbles, for example, have now been found that operate at less than one meter per second.

More elaborate techniques have been developed, too. They include: watching for long-term changes in the radio signals that arrive at Earth from a pulsar (an aged and rapidly rotating star), since this might indicate a second star and perhaps also a planet that is orbiting this binary (two-star pair); detecting the small amount of heat emitted by an exoplanet itself; and analyzing the atmosphere and composition of exoplanets. Though astronomers are working toward the goal of actually photographing more exoplanets (as opposed to seeing the effects of their existence on the stars), so far only two have actually been seen visually, and these may be brown dwarfs rather than true planets.

> A planet with three Suns to call of its own has been found. This exoplanet has a main Sun, which is large and yellow, and also very close; the planet orbits it in fewer than 4 days. Farther away, at about 10 AU, shines a medium-sized orangeSun, along with a smaller red one. Beautiful sunrises and sunsets might become almost too distracting on this world.

Exoplanets in the Making?

Elsewhere in the universe, planetary inhabitants (if there are any) would never choose to live around OTS 44. This object is a brown dwarf, a failed star yet even it has a nebula around it in which planets could form and, indeed, may now actually be forming. Instead of a healthy star emitting significant light and heat, OTS 44 is only about 15 times the size of Jupiter and has a temperature of about 2300 degrees Kelvin. Any livable planet in this situation livable by our standards at least would have to huddle quite close to OTS 44 or suffer cold comfort.

The two reasons that this kind of unusual astronomical situation is important and not merely interesting are: if a disk of gas and debris can form around an object of this kind, merely a brown dwarf, there are probably many more exoplanetary disks and of differing masses than astronomers before imagined; it may be easier to spot planets around brown dwarfs than around stars since the dimmer light from a brown dwarf is less likely to swamp out the light of planet.

Another frontier in exoplanet analysis involves the possibility that planets could indeed form around typical binary stars. Though it was once thought that two stars would devour all the protoplanetary material, or at least interfere through gravity with its consolidation, astronomers no longer think so. Instead, all the extra tugging actually helps the loose material to consolidate into planets. This issue is important since about two-thirds of the stars in our galaxy are binaries or in even larger immediate groupings.

Unusual Combinations

In the spring of 2005, astronomers found another unusual pair of objects. One of them is a star named GQ Lupi. The other, an object far from that star at about 60–130 AU, may be a planet, or it may be a brown dwarf. If it is a single planet, it should not be that far from its star with no other planets in between, since a solar nebula's disk consistently has material at distances going outward. Perhaps there are forces such as a companion star, which flung this possible exoplanet out. Or there are more smaller exoplanets in between still impossible for us to detect. These "gas planets" might be more like the Earth.

Yet another odd scenario was discovered a bit before this, in late 2004: a trio of objects. They might be a single exoplanet orbiting both a star and a brown dwarf, or the proposed brown dwarf may instead be the most massive planet ever found, and the original planetary disk may also be much heavier than any that have been identified.

The lightest mass exoplanet discovered so far, in late 2005, is only about 7.5 times more massive than the Earth. It is orbiting a red-dwarf star, Gliese 876, only 15 light-years away. The planet lies very close to its star, at a distance of only 1,922,000 miles (3.1 million) kilometers.

> The most Earth-like exoplanet found so far, announced in 2007, lies about 21,000 light-years from here. It is rocky, about 5.5 times as massive as the Earth and a bit smaller than Neptune. Called OGLE-2005-BLG-390Lb, this planet is about 234 million miles from its star, a dim reddish denizen of our galaxy. Chilly but somehow familiar.

Solar System Look-Alikes?

Not only solar systems in the making but mature solar systems have also been found. One of these, around a red dwarf star a very old star named AU Microscopii has the leftovers of a planetary disk called a debris disk. So does another star, the younger HD 107146, a middle-aged one named Fomalhaut, and a Sun-like star named HIP 8920. The leftover disk around the latter star is at about the same distance from its star as we are from the Sun. These debris disks are similar to debris that formed the asteroid belt in our solar system and the reservoir of comets and other objects in the outer reaches. (These are discussed in Chapter 7, The Smaller Bodies of Our Solar System.)

Debris disks, which form much later than the original disks such as our solar nebula, are more diffused and of lower mass. They include leftover material from the original disks as well as debris from the collisions among bodies that formed previously. Since these collisions remain quite frequent in the early days of a solar system, debris disks can at least be replenished for several million years. These disks around other stars could, then, easily hide small, Earth-sized planets.

Exoplanets: Next Steps

Astronomers are of course interested in the natural history of the whole universe. But finding a solar system truly like our own cannot help but be a major, if parochial, goal. What might that look like and what are some of the discoveries necessary on the way to finding it?

The first needed are techniques to identify smaller exoplanets, preferably rocky ones, and at a distance from the central star that would make the planet's surface neither too hot nor too cold. (Some astronomers call this the Goldilocks issue, since the porridge she sought was to be neither too hot nor too cold.) So far, the exoplanets found, which are anywhere near the mass of Earth ones at 7.5 times, 10 times, 14 times, and 20 times our size are far too close to their stars. Close-in objects are easier to spot, but they are not hospitable to life as we know it. Exoplanets near to our mass, however, are at least probably terrestrial rocky and solid instead of being gas giants.

A second goal is to find an exoplanetary system closer to our own neighborhood. One has now been detected at only 15 light-years away. But the object there most like the Earth is far too hot.

And a third step would be to find a system where the planets lie in circular or near-circular orbits. If the outer planets in a solar system orbit in this fashion, the prognosis for any inner planets is better in the sense of safer. They are likely to orbit along similar paths and be less likely to be pulled off course into collisions, or tossed out of the system into far-flung oblivion. So far, just one such solar system has been found around the Sun-like star HD 70642 about 90 light-years away from here.

Astronomers also want to analyze the composition of the atmosphere of exoplanets found. How much water vapor is there? Is there oxygen? Details like this would be useful to know.

In this chapter you have learned that:

✔ Our solar system and other solar systems are formed out of a solar nebula of gas, dust, and ice.

✔ The Sun formed first, from the hot, churning center of the disk.

✔ The rocky planets were formed in the inner regions where rock and metal particles collided and melted together to make larger and larger objects.

✔ The gas giant, or Jovian planets, were formed in the colder regions where ice cannot melt.

Jupiter, the largest planet in our solar system, would have become a star if it had been 80–90 times more massive.

 Saturn, host to a recent Cassini spacecraft visit, is becoming slightly less mysterious.

 Uranus spins on an axis at almost right degree angles to the orientation of all the other planets.

Neptune, though very far from the Sun, still has seasons.

Exoplanets are being discovered regularly; though most of the ones detected are very much larger than the Earth (some that are close to our size have been detected, though they orbit so close to their stars that their surfaces would be scorchingly hot).

Test Yourself

True/False Questions

1. An interstellar cloud is full of dark matter.
2. The Jovian planets are all gas giant planets.
3. The solar wind blew away most of the leftover cloud material by the time the solar system was 4.6 billion years old.
4. Jupiter has rings.
5. Jupiter may be a brown dwarf.
6. Saturn has four solid rings.
7. Neptune tilts sideways on its axis.
8. The exoplanets found so far suggest that our type of solar system is common.
9. It is just as likely to find large exoplanets as it is to find Earth-sized ones.
10. An exoplanet has been found that has three Suns in its sky.

Completion Questions

1. The solar nebula was primarily made of two gases, _____ and _____ .
2. The Sun is about _____ billion years old.
3. The Sun burns its hydrogen/helium fuel at about _____ degree Fahrenheit.
4. Neptune's distinctive color is _____ .
5. About _____ exoplanets have been discovered so far (as of 2006).
6. The youngest exoplanet discovered so far is about _____ years old.
7. The oldest exoplanet discovered so far is about _____ years old.

8. The exoplanet discovered so far that is closest to the mass of the Earth is about _____ light-years away.

9. A solar system where the planets are in near-circular orbit is that of a star _____ light-years away.

Multiple Choice Questions

1. A solar nebula is
   a. A cloud around the Sun today
   b. A gas and dust area where our next Sun will be born
   c. The cloud out of which the Sun and planets formed

2. The core accretion model of planet formation posits that
   a. Concrete is at the core of each planet.
   b. The planets were built up by collisions in which more and more material stuck together.
   c. All the planets have cores.

3. An exoplanet is
   a. A planet found outside of the solar system
   b. A planet once in our solar system but pulled away by a star's gravity
   c. A planet that is the largest in its solar system

4. Exoplanets have so far been discovered primarily by detecting
   a. The wobble of their star
   b. Their light
   c. Their relationship to other planets

5. A debris disk is
   a. The leftover material around the Sun from crashed comets
   b. A place to look for more exoplanets
   c. Material that lies between the orbits of Mars and Jupiter

ANSWER KEY

True/False
1. F   2. T   3. F   4. T   5. F   6. F   7. F   8. F   9. F   10. T

Completion Questions
1. Hydrogen and helium   2. 5   3. 10,000   4. Blue   5. 200   6. 1 million
7. 12.5 billion   8. 15   9. 90

Multiple Choice Questions
1. c   2. b   3. a   4. a   5. b

CHAPTER 6

The Solar Nebula and the Formation of the Rocky Planets and Moons

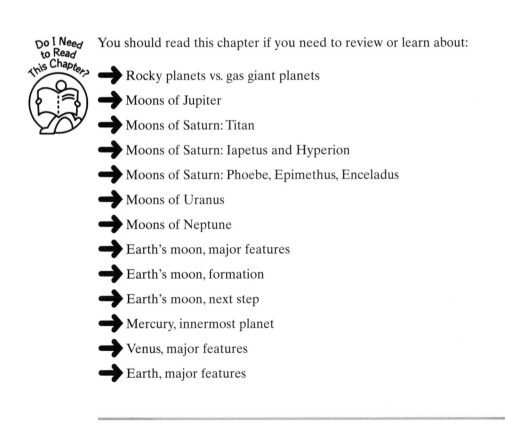

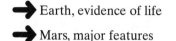 Earth, evidence of life

 Mars, major features

 Mars, the water issue

 Mars, the life issue

 Moons of Mars

 Moons beyond/moons vs. planets

Get Started

 Our solar system, and presumably other solar systems, has an immense variety of rocky moons along with its smaller number of rocky planets. Though the planets of the Sun probably formed in essentially the same orbit they occupy now, some of the moons in our solar system were captured from elsewhere.

Rocky Planets vs. Gas Giant Planets

In the swirling cloud of gases and dust that was our original solar nebula, the inner areas were more dense and warm than the outer ones. The solid particles scattered there were very numerous and initially small, generally less than a centimeter in diameter, but they soon began to collide with considerable force. So, over time, grew the rocky planets Mercury, Venus, Earth, and Mars their gravity pulling in additional materials as long as it was available near to them. This scenario is in contrast to the more outlying regions of the nebula where the gas giants Jupiter, Saturn, Uranus, and Neptune formed, as described in Chapter 5, The Solar Nebula and the Formation of Gas Planets. There, it was colder cold enough that ice could condense in the nebula and the colder gases could be drawn in by gravity. The growing gravity of these large planets drew in the diffuse hydrogen and helium gases in immense amounts. They are formed primarily of them today.

Moons of Jupiter

This most immense planet, Jupiter, also has the most moons about 60 and all are ice and rock objects. At least 53 of Jupiter's moons are known to be irregulars, meaning that they do not orbit Jupiter in a regular near-circle orbit in the same plane as the planet itself and in the same direction as the planet rotates. All small, they probably are leftover rocky bodies from the early era of the solar nebula. Some even seem to be captured asteroids. Jupiter's gravity is strong enough to pull in anything near to it, and the early solar system was a quite chaotic place.

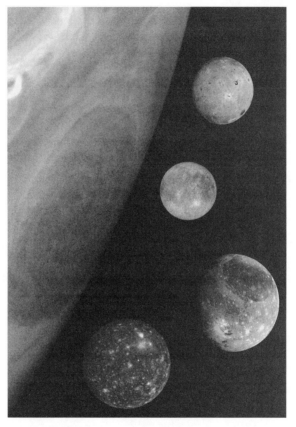

Montage of Jupiter and its four largest moons, the Galilean Satellites. NASA

Jupiter's other, "regular," moons number about seven. These formed in the same way the solar system's inner planets did, but out of a smaller part of the solar nebula the part held nearby and orbiting Jupiter as it itself formed. The best known of these are the four largest moons of Jupiter Ganymede, Io, Europa, and Callisto.

Ganymede is our solar system's largest moon. With 3274 miles (5269 km) in diameter, it is larger than the planet Mercury.

Io, a bit smaller than Mercury but larger than Pluto, is known for having active volcanoes, more than 80 of them. Unlike the other three largest moons of Jupiter, it is mostly rock. Red and orange patches, of sulfurous exudations of the volcanoes, blotch its surface. This vulcanism is probably caused by the strength of Jupiter's gravity, which squeezes Io's rocky core in a way that resembles ocean tides on Earth except that force acts on solid rock and metal. Tidal forces like these can indeed create enough heat, and thus molten material, to make volcanoes erupt.

Europa probably has a liquid water ocean under an icy lid, while Callisto and Ganymede are mostly ice and rock. All of these moons are even larger than Pluto.

Quick Tips

1. The four large moons of Jupiter are known as the Galilean moons since they were discovered by Galileo using the telescope he invented.

2. Io sometimes carries the nickname Pizza Moon because of its colorful splotchy surface.

Image of Io, the innermost of the Galilean Satellites. The peculiar colors of the surface of Io are due to compounds thrown out by Io's many active volcanoes. The inside of Io has been liquefied by the heat generated by tidal friction with Jupiter and Europa. NASA

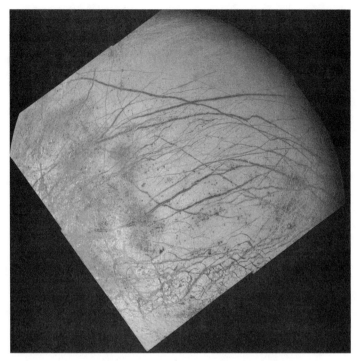

The surface of Europa is relatively smooth, soft water ice. There are cracks and other features that can be seen in this image. NASA

Moons of Saturn: Titan

If astronomers were polled in a popularity contest, Saturn's moon Titan would probably win as the most popular moon in all of the solar system. The reasons are: it is the most interesting, and it has some similarities to the Earth. Titan has a significant atmosphere, made of nitrogen, carbon, hydrogen, hydrogen cyanide, argon, propane, acetylene, ethylene, ethane, methane, and other hydrocarbons, colored a murky orange. (This makes it distinctly unbreathable for humans.) The extent of Titan's atmosphere was discovered several years ago as the moon passed in front of the Crab nebula from our vantage point on Earth; Titan's atmosphere absorbed some of the x-rays emitted by the supernova, or exploded star, remnant that makes up the Crab.

Now that the Cassini spacecraft has recently made its 50 separate pass-bys of the Saturnian system, with one probe also sent down to the surface through Titan's atmosphere, much more is coming to be known. Titan's surface, coated with organic hydrocarbons, include liquid pools and lakes, some of them as large as 90 kilometers across, made of methane and perhaps ethane. In between are drier features, which look like drainage channels (up to 43.7 yards or 40 meters across), hills, ice volcanoes, and sandbars, all punctuated with river rocks that seem to be coated with water ice. Frozen water appears to form a great deal of the surface area between the methane pools and lakes, though more of the liquid methane seems to be below the frozen water. And methane clouds above complete a hydrological cycle like the Earth's, though primarily with methane instead of water.

It is in extreme and oxygen-free environments, like that of Titan, that the first life may have developed on our own planet. Called extremophiles on Earth, versions of these ancient bacteria-like creatures could possibly also proliferate in such a cold and poisonous place as Titan.

Titan is geologically active, too. Scientists posit this since no craters are visible, even though every moon in the solar system is typically hit regularly by meteoroids and other space debris. Since Titan lacks a magnetic field (a kind of shield), it is also blasted by cosmic rays regularly, some of which material surely penetrates its thick atmosphere. So something major is remodeling the surface of this moon, to make it smoother rather than pock-marked by craters. Whether this is methane, water ice, ammonia, volcanoes, snowfalls, methane rainfalls, or something else is not known.

Montage of Saturn and its many large moons. NASA

A remarkable look down through the thick clouds of Titan, the largest moon of Saturn, revealing the never before seen surface of this mysterious world. NASA

Titan is the second largest moon in the solar system, after Jupiter's Ganymede. It is a bit bigger than the planet Mercury.

Quick Tip

Some of Saturn's 30-some smaller moons (all of which are rocky like its major moons), act as "shepherds," corralling smaller particles within the rings into clusters, which then orbit along with them, like eddies in a flowing river. More information about its moons and rings is found in Chapter 5, The Solar Nebula and the Formation of Gas Planets.

Moons of Saturn: Iapetus, Hyperion, Phoebe, Epimethus, and Enceladus

Though Titan is a hard act to follow; Iapetus also stands out among Saturn's moons. It features extremely lofty, white mountains—at 6–12 miles (9.65–19.3 kilometers) high, they may be the highest anywhere in the solar system. Even stranger, this moon is almost completely black on one side and white on the other.

Phoebe is known as an irregular moon, probably captured by Saturn's gravity from the vast halo of frozen bodies beyond Pluto called the Kuiper belt. (See Chapter 7, The Smaller Bodies of Our Solar System, for more on that zone of our solar system.) Astronomers had always known that Phoebe was irregular— it orbits in the opposite direction from Saturn's own rotation, which means that it did not form out of the early swirl of material orbiting Saturn. And its orbit swings high and low above and beyond the plane of the planet's rings. The Cassini spacecraft has shown that its particular composition and density of rock and ice makes it similar to Kuiper belt objects and unlike the other Saturnian moons. Images also reveal plenty of craters, ridges, landslides, and steep channels on Phoebe.

Epimethus is small and not at all round. Made largely of water ice, it has been pummeled hard by meteorites, which have left craters and even knocked off part of its body, as though a boxer had been hitting it until it became misshapen.

Enceladus, also tiny with a surface area smaller than the state of New Mexico, is becoming more of a focus of astronomers' attention than before, since an eruption of water vapor, dust, and tiny ice crystals has recently been seen emerging from an area of cracked surface on the moon. Within the emitted material are both inorganic and organic compounds, and perhaps amino acids—the building blocks of life—(live in the below-surface liquid. It has an atmosphere of water vapor, with lesser amounts of molecular hydrogen, carbon dioxide, carbon monoxide, and molecular nitrogen. This atmosphere, though thin, extends about as far out into space as this small moon's diameter. It is coming from the cracks and perhaps also from exhaled surface vapor, geysers, or ice-erupting volcanoes. (The material is not held securely near the surface, since gravity is so weak here.) Volcanic eruptions are very common on Enceladus. The heat source is gravitational flexing by Saturn, leading to tidal forces in the ice.

With this moon's gravity so weak, it is constantly losing part of the atmosphere, which is then replenished. The lost material, much of it particles of water ice, becomes part of Saturn's rings, and, in fact, meteoritic dust from Enceladus feeds one of the rings constantly.

Image of Dione, an icy moon of Saturn with the planet and its rings in the background. NASA

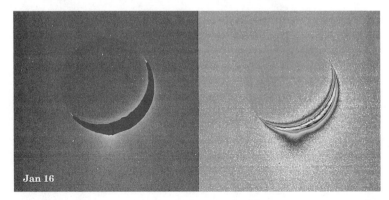

Contrast enhanced images of Enceladus showing water spray erupting from fissures in the surface of this moon of Saturn. NASA

Enceladus has a diameter of only 314 miles (505 kilometers), it may have a subsurface ocean, and it has a surprising abundance of surface features from faults to ridges to folds to stripes (though no craters). Its surface is also shiny.

But this moon is large compared to two tiny moons discovered near to it. Called S/200451 and S/200452 (at least for now), these are 1.86 miles (3 kilometers) and 2.5 miles (4 kilometers) in diameter, respectively. A morning's hike on these three would not be tiring, though temperatures of 80–110 degrees Kelvin (provided only by heat seeping up through cracks on its surfaces as a result of former tidal flexing) would make it extremely chilly.

Moons of Uranus

In addition to the material orbiting it within its faint rings, Uranus has at least 21 moons. The largest are Titania, Oberon, Umbriel, Ariel, and Miranda, made mostly of iced ammonia, methane, and water, though with rocky cores. Little Miranda shows evidence of considerable tectonic activity. The heating of its core to provide the engine for this geological activity might have come from the tidal forces like those described above; or it is possible that Mirando suffered a massive collision with another object in the past.

Titania is a moon of Uranus. Mostly made of water ice, Titania and most of the moons of Uranus have a dirty, dusty surface. NASA

Partial mosaic image of Miranda, a small, odd moon of Uranus. Miranda looks as though it was partially broken apart at some time in the past and then reassembled itself into the patchwork of geological features we see today. NASA

Moons of Neptune

This bright blue planet sweeps around the Sun with at least 11 moons for company. The newest two discovered (in 2003) are small, both in the range of 18–25 miles (29–40 kilometers) in diameter. They orbit Neptune in a plane different from the rest of the solar system. This irregularity suggests that they were captured in the early days of the solar system or were created in a collision at about that time. The small pair is also about 60 times farther from their planet than is Triton, the largest moon. Triton may well have been a planet like Pluto once.

Triton, bigger than Pluto, is also an irregular, orbiting retrograde (in this case, clockwise, rather than counterclockwise) around Neptune. Made of rock and ice, it has ice volcanoes and ice geysers. Present also are dry ice and frozen hydrocarbons from methane and nitrogen to carbon monoxide. This moon has an atmosphere, but a very thin one, probably from the regular sublimation of ices on its surface.

Composite image of Triton and the planet Neptune behind in the distance.
Triton is a cold, bizarre world of rock, ice and liquid gasses. NASA

A piece of the Earth's Moon was found, on Earth, in early 2005. Small at about 2.75
inches (7 centimeters) across, black, and named Sayh al Uhaymir 169, this rock was
discovered in the desert of Oman. It had been lying there for about 9700 years.

Scientists could tell that this rock was not of our planet. It was not magnetic, was
radioactive, and had been thoroughly blasted by cosmic rays in space. This information
established its origin: the Moon's Mare Ibrium basin. This basin formed in a giant
meteorite hit long ago, followed by three other impacts from meteorites over time. The
fourth hit—about 340,000 years ago—had blasted this very rock (and certainly others)
away from the sway of the Moon's gravity and out into space. There, it orbited the
Earth, subject to the intense cosmic rays above our atmosphere, until it fell down to
the surface. The date was established by looking at the beryllium and carbon isotopes
in the rock, which stopped changing when the cosmic rays ceased their hitting of the
rock. That was when it had fallen to Earth: a "messenger" from the Moon.

Earth's Moon, Major Features

Our planet is unusual in the solar system for having a moon so close to the size of the planet it orbits. The Moon, at 2160 miles (3476 kilometers) in diameter, is about one-fourth the size of the Earth. A solid rock, it has no real atmosphere, just wispy bits of hydrogen, helium, neon, and argon, probably from rocks vaporized upon impact. These atoms escape regularly and readily into space.

The Moon is scarred with craters. The impacts that created them came from meteoroids, comets, and their debris that pummeled its surface, since there was, and is, no appreciable atmosphere to slow them down. Some of the craters are tiny pockmarks, others hundreds of kilometers across. The Moon's largest crater is more than 1500 miles (2414 kilometers) in diameter. Many of these impacts occurred in the early days of the solar system when local space was cluttered with remaining debris from the original solar nebula. Their evidence remains here because nothing has blotted it out. The Moon has not, in ages, had the tectonic processes to do so—no mountains to rise up, nor volcanoes to flow with lava to fill in the craters, nor any crustal movements. Some small volcanic cones do remain in place, however, from when the Moon's core was hot; that was long ago, a brief time during which it could power tectonics.

The Moon orbits and also rotates counterclockwise, just as the Earth does around the Sun, and it lies in very close to the same plane as we do. Its rotation is also synchronous with ours—it spins on its axis one turn just as it makes one

The Moon, our Earth's very own satellite. The Moon is about as large as the moons around the gas giants and thus is unusual in that it orbits a rather small planet by comparison, the Earth. NASA

Astronaut Buzz Aldrin standing on the Moon. The Moon is the only Solar System body other than the Earth that humans have set foot on. NASA

Star parties do not involve earthly celebrities. Rather they are all-night, weekend (or longer) parties at which hundreds of people gather to observe the stars. Many participants have telescopes and readily allow those who do not to look through theirs.

orbit around us. For this reason, we see only one side of it, from Earth, and always the same side.

These events are always set outdoors in places far from city lights, during seasons when clouds are least likely, which even away from any mountains, so that the whole dome of the sky is brightly visible. Star parties are always at the time of the New Moon, when the Moon is not visible, which makes it even darker. Partygoers can often thereby see 2000–3000 stars with the naked eye, and far more by binoculars or telescope.

To find out when and where upcoming star parties are planned, go to www.skyandtelescope.com and find their event calendar, which has a list of upcoming star parties viewable by state. They tend to be quite inexpensive.

The Moon's main features are:

- Highlands, old rocks which are densely cratered
- Maria, old, smooth, and dark volcanic plains
- Impact craters, where space debris punched in
- Mountains, higher areas which were thrown up by the impacts
- Rays, sandy material thrown out in the impacts
- Rilles, now-solid channels formed by the Moon's ancient, dead volcanoes
- A few astronaut footprints in the sand

Quick Tip

The Moon, with about one-fourth the diameter of the Earth, has the highest such ratio in the solar system.

Earth's Moon, Formation

Though people have puzzled for eons about how the Moon came to join us in space, so close and so beautiful, astronomers believe that they have finally figured it out. The pairing happened long ago—about 4.5 billion years ago in the very infancy of the solar system—and it involved a great deal of violence in our own neighborhood. It is the Moon's composition that tells us it formed near to us in the original solar nebula.

What happened? A "rock," about the size that Mars is today, smacked into our new planet. (Anything much bigger than this would have broken us into pieces.) This massive collision catapulted plenty of debris from the Earth into near-space, and it remained there, under the control of the Earth's gravity. Smaller collisions among these chunks gradually melded them together, followed by larger chunks in larger collisions. Thus formed the Moon, large and already in place. The birth seems to have been between 4.537 and 4.517 billion years ago, a 20-million-year period when the solar system was only 30–50 million years old.

Quick Tip

The Moon is made from material that looks very similar to the Earth's crust and upper mantle. And this is indeed the material, blasted off, which formed our companion.

An unusual image of the Earth and the Moon, taken from a
receding spacecraft. Notice how blue the Earth looks in comparison to the Moon, which has no air or water. NASA

Apollo spacecraft have visited the Moon nine times, from the first two fly arounds in
the late 1960s up to the most recent human visit by astronauts in late 1972. An additional, tenth spacecraft, robotic, orbited the Moon in 2004. Some of these explorations
resulted in Moon rocks being returned to Earth for study. Much of what we know
about the Moon has been gleaned from laboratory analysis of these rocks.

A puzzle of the Moon's formation remains, however. How did an immense
Mars-sized chunk such as the one that hit us could have been drawn off its orbit
to make that initial collision possible? After all, Venus, Mars, and other planets
stay in place, in their own orbits. The answer seems to be that this giant "rock"
formed and was occupying one of the two zones where the gravity of the Sun
and the gravity of the Earth are essentially equally strong. The collision object
probably remained there for awhile. Then, in our young solar system with much
material still loose and playing "bumper cars," something reasonably large
nudged it just slightly out of this zone. It was pulled toward the Earth by our
gravity, the massive collision occurred, and the Moon was born.

Earth's Moon, Next Step

Investigation of the Moon continues, from both the spacecraft recently in orbit
and from our vantage point on Earth. Astronomers wonder how much water
vapor exists there, search for ice in the shadowy rings of its polar craters, and ponder where a Moon base might be situated. The future will almost surely bring
another trip, though none is now scheduled.

Mercury, the Innermost Planet

Mercury looks like Earth's Moon, cratered and with no appreciable atmosphere. It is not even much bigger than that body, 3031 miles (4878 kilometers) in diameter vs. the Moon's 2160 miles (3476 kilometers). This makes Mercury's less than half the diameter of the Earth. Mercury is also a bit larger than Pluto (if that body is counted among the planets, a controversy described in the last section of this chapter).

It is truly a small—but very hot—world in contrast to the Moon's and Pluto's chill. Mercury is so close to the Sun that its noontime temperature rises as high as 700 degrees Kelvin. No atmosphere cloaks its transition to nighttime. At sunset, temperatures quickly drop to as low as 100 degrees Kelvin (300 degrees

Mercury is a desolate planet with no significant atmosphere or surface activity. Its heavily cratered surface resembles the surface of the Moon. NASA

Fahrenheit). At its poles, though, some ice consistently lies below the surface's dust, sand, and rubble. This planet's average distance from the Sun is 36 million miles (58 million kilometers), huddled close to make all that warmth.

> Mercury has no moon of its own. A planet that close to the Sun probably had any such chunks of matter pulled straight into the Sun, long ago.

Mercury's orbit is not in a flat plane; in fact, it is inclined about 7 degrees from it (more than any other planet's except Pluto). This tilted path takes it as far from the Sun as 43.5 million miles (70 million kilometers) and as close as 28.6 million miles (46 million kilometers), depending upon its position in its orbit.

Core and Formation

Mercury's core is unusually large for the size of the planet as a whole. This highly metallic core comprises most of its body. Just a 435 mile (700 kilometer) thickness of rocky mantle and crust surrounds a core about 2175 miles (3500 kilometers) in diameter.

> The closer a planet is to its Sun, the shorter is its orbital period and the faster is its speed around this orbit. (This is from Kepler's third law of motion.) Mercury whizzes around the Sun in 88 days, at a speed of 29.8 miles (48 kilometers) per second.

The unusual composition of Mercury indicates something about its probable history. Scientists believe that it suffered at least one immense impact in the early days of the solar system, one that knocked much of its outer layer material back off into the solar nebula. After that, the planet was battered by all the space debris, so common in the early days of the solar system, yielding the cratered surface visible now.

Quick Tip

This innermost planet is as atypical within our solar system as is Pluto, the outermost object in the inner solar system. Both are close only to the size of a large moon, just as Earth's Moon; and both traverse orbits highly inclined to the plane in which the other planets move.

Venus, Major Features

Lying between the Earth and Mercury, Venus is our sister planet, but with a twist. Though near to us in size—7520 miles (12,102 kilometers) in diameter versus Earth's 7926 miles (12,756 kilometers)—and not too much closer to the Sun, Venus has an inferno-like surface temperature of 870 degrees Fahrenheit. This heat is held high and nearly constant by the planet's thick cloud swaddling, most of it carbon dioxide, with nitrogen gas, and sulfuric acid droplets. The thick atmosphere creates a reddish-orange fog, which never lifts, and also smells bad. Even if Venus had a moon, it would not be visible to anyone looking up from the planet's surface.

If a visitor were able to survive the Venusian heat, the barometric pressure would still be drastically oppressive here; it impinges on the surface at about 93 times the force of the barometric pressure on Earth.

Venus' surface features, all rocky, include: plains, mountains, volcanic flats, volcanic domes, canyons, and rilles. These features can be detected from Earth, but only by cloud-penetrating radar. Craters from the meteoritic impacts in the early days of the solar system have been mostly blotted out by the volcanic activity. Since then, the thick, murky atmosphere has also incinerated a lot of the smaller incoming material, though no plate tectonic activity occurs to help smooth the planet to any greater degree. Vulcanism probably does continue on Venus today; a giant spurt of this activity occurred as recently as about 500 million years ago.

Venus does look beautiful from the vantage point of Earth, however. Since it is close to us, it is usually the brightest object visible in our night sky. (Only Jupiter, farther away but much larger, outshines it at times.) Since Venus' brightness varies, depending upon how close we are together in our differing orbital swings, it can appear surprisingly bright to inexperienced observers on Earth. People occasionally call astronomers to report a Unidentified Flying Object (UFO).

Because of the uniformly blanketing atmosphere, Venus seems to have little wind. Its hardened basalt from the volcanic eruptions lies relatively undisturbed. No rain ever falls down as far as the surface, though sulfuric acid rainfall occurs within the clouds.

Venus Facts

- With a close-to-circular orbit, Venus is usually about 67 million miles (108 million kilometers) from the Sun.
- Its year is about two-thirds as long as ours.
- Venus rotates very slowly on its axis, in 2.243 days, as opposed to our nearly 24 hours; so its daylight hours last long, as does its night.

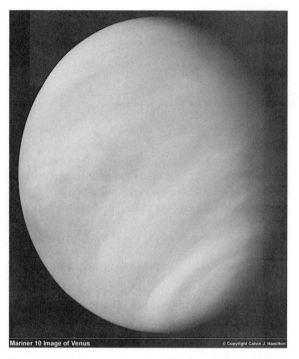

Venus is enshrouded with a very thick atmosphere. Here we see only the very top of the clouds swirling around the planet. NASA

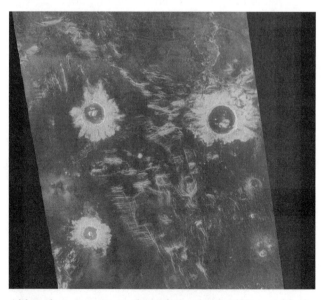

Although we can not see through the clouds of Venus with visible light, radar can. This radar image shows impact craters on the surface of Venus. Even though Venus has a thick atmosphere, large objects wandering around the Solar System can still occasionally hit the planet and make a crater. NASA

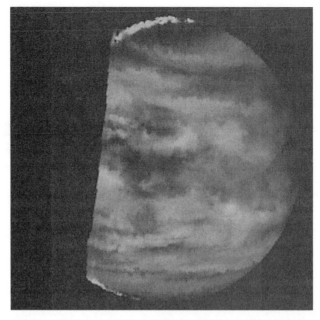

An infrared image of the night side of Venus. The features
show different temperatures between the clouds blanketing
Venus. NASA

A computer generated mosaic of radar images of Venus
made by the Magellan spacecraft. NASA

This amazing picture was taken on the surface of Venus by the Russian Venera 13 probe. Imagine surviving a flight through an atmosphere with sulfuric acid droplets to land on a surface with a temperature hot enough to melt lead and an atmosphere 100 times denser than here on Earth. NASA

Quick Tip

Venus is a much more static planet than Earth. The weather hardly changes, the daytime temperature barely stirs, and no major plate tectonics pushes up mountains or remodels shorelines. In fact, there are no shorelines—no oceans, lakes, or rivers. Volcanoes, however, do occasionally erupt.

Earth, Major Features

To an observer arriving from another planet, the Earth would look varied in color and topography. About 75 percent blue (the waters of the oceans and lakes), about 2 percent white (the ice of the glaciers, mostly at the poles), and about 23 percent dressed in greens and browns, with white clouds afloat, all this would complete the picture. We would be considered a medium-sized planet within our solar system, with a proportionally large moon and considered a planet in a stable solar system at a medium distance from our star.

Closer in, the geologic array would become quite evident: a magnetosphere to cloak us, along with an atmosphere; then surface mountains, volcanoes, and plate tectonic movements of all kinds resulting in continents; on these continents, low plains, canyons, deserts, and more. Also, a profusion of carbon-based life forms would be detectable. Below the surface, a molten mantle (down to approximately 1864 miles or 3000 kilometers below the surface) and a solid core would be noted. The core's extent, about 4350 miles (7000 kilometers) in diameter, would indicate a significantly dense planet, and one just under 8078 miles (13,000 kilometers) in diameter as a whole.

Worth a Visit?

Would the Earth appear to be an interesting place to visit (or does even that question reflect a ludicrously parochial perspective)? Earth is indeed the planet in this solar system with the most variety—of topography, of atmosphere, and of seasonal change, at the very least.

But that may or may not be saying much when one considers the universe, or even the galaxy as a whole. It probably depends where you come from. We have, after all, only one Sun and one moon to enliven the view, day and night. There are visible stars, but they can be seen from almost everywhere else in the galaxy anyway.

An image of Mt. Everest, the tallest mountain on the Earth, from space. Notice the effects of water erosion on the mighty Himalaya Mountains, carving large valleys and sculpting the landscape. Erosion and plate tectonics erase evidence of most impact craters that occasionally form when a large object smashes into the Earth. NASA

This image from orbit shows hurricane Isador in the Gulf of Mexico. The Earth has a very active atmosphere with complex winds and clouds. NASA

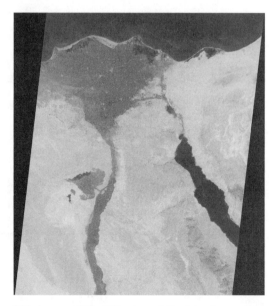

This image of the Nile river delta take from space shows the darker regions with vegetation watered by the river. Any alien flying by the Earth would immediately be able to tell there was life on this planet. NASA

Earth, Evidence of Life

Observers contemplating a visit would probably wonder about the existence of living things here. Some favorable conditions would be immediately evident: air, liquid water, a solid surface, the variety of surface features, a planet with a stable rotation and a stable orbit, not too close or too far away from the star; and a magnetosphere to protect the surface from a deluge of cosmic rays.

Might there be intelligent life here? Or merely some interesting microbes and a few leftover dinosaurs with miniscule brains? Observers could surely intercept the radio waves and other electronic emanations from this planet. They could see the electric lights at night. Probably communication with something down there is possible, they might conclude. A visit? That's a big maybe.

An unusual image taken from the Space Shuttle while in orbit. The glow in the background is Earth's Aurora see along the limb of the Earth. NASA

Mars, Major Features

Mars, like Venus and Earth, is a medium-sized, rocky planet. It has a core of liquid and/or molten iron and was once a tectonically active body. Vulcanism and perhaps plate tectonics have made here high volcanic plains, layered volcanic rocks, crustal cracks that look like canyons (one slices a line almost a fifth of the way across the whole planet), low sand and gravel plains, rock outcroppings, mountains, and volcanic domes larger than any on Earth. Some of the volcanoes have been active within the last 2 million years or so and are thus not considered dead.

Mars does not have a strong magnetic field the way the Earth does, but it does have some remnant magnetic regions. Because of them, Mars has faint Northern Lights, or aurora.

Its predominant surface color—reddish-orange—also signals that this is an iron-rich planet. The place looks like a rubble desert in parts of the western United States, though drenched in a deep sunset light and with many extra craters from meteorite impacts.

The atmosphere on Mars is very thin—about one percent the density of the Earth's. It is mostly carbon dioxide gas with faint traces of nitrogen and oxygen (compared to Earth). Winds, though, are fierce, kicking up violent sandstorms. Temperatures are cold, averaging -76 degree Fahrenheit.

More Martian facts:

- Diameter, 4219 miles (6790 kilometers), a bit more than half of the size of the Earth's and also smaller than Venus
- Tallest volcanic peak, Olympus Mons, rises 15.5 miles (25 kilometers) above its surroundings (it may well be the largest volcano anywhere in the solar system; its base alone would cover all of Arizona)
- Largest crater, 77–93 miles (125–150 kilometers) across
- Most recent crater, 22 yards (20 meters) across, only about 25 years old
- Average distance from the Sun, 1.52 AU (half again as far away as we are)
- North and south pole ice caps made of dry ice which change with the seasons
- Length of day, almost 24 hours

Quick Tips

1. Mars is smaller than the Earth and Venus, larger than Mercury and Pluto.
2. The planet is rich in iron minerals, which exist in its core and can also be seen on the surface.

This image shows the full disk of Mars as seen by the Hubble Space. Notice that Mars has a polar ice cap at the South pole. NASA

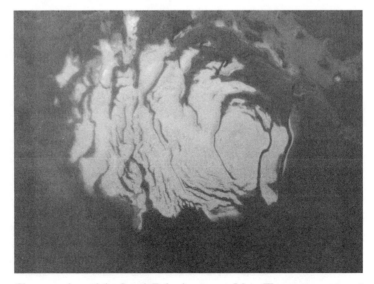

Close-up view of the South Polar ice cap on Mars. The permanent part of the ice cap is water, not frozen carbon dioxide. NASA

The remains of an old volcanic caldera, Apollinaris Patera, on Mars. This image captures a thin patch of bright clouds over the summit. NASA

Ophir Chasma, a part of the huge Valles Marineris canyon system. Notice how the steep cliffs have broken loose and slid downhill into the lower valley. NASA

Mars, the Water Issue

Finding water on another planet would be quite significant to us, since water is the basis of life as we know it. The surface of Mars features gullies (some with sand dunes), channels, rilles, gorges, rocks with dried sulfur salt deposits, smooth rounded rocks, deep ripples (with sides taller than a grade-schooler), riverbeds, and other topography which looks as though it was created by flowing water, melting snow, and falling snow. Among its surface minerals are also those known to have been altered by water. These include clays, hydrated sulfates, iron oxide, goethite, and jarosite. Astronomers believe that the case for the existence of water, or at least water ice, on Mars is very strong.

However, it does not look as though this planet was ever home to long-term, large, warm oceans. Water probably existed here in two periods, a longer one that ended about 3.8 billion years ago (and probably lasted for a few million up to 500 million years) and another, briefer and more recent period of more acidic water. One mineral has been found to be common here—olivine—which is not stable in the presence of water, and there is enough of this olivine to indicate that the water was not widespread. Only small amounts of carbonate minerals have been detected, the formation of which could have been blocked by all the vulcanism, rather than being limited by the amount of water.

> NASA has recently sent three orbiting and two robotic surface-rover spacecraft to Mars, exploring two vastly different regions of the planet's surface. All of the information on Mars, beyond the basic statistics, was gathered by these tiny solar-powered jeep-like vehicles. They beam it back to Earth.

What happened to the water that was once here? Was it flood water, which then soaked underground, where it remains hidden today? Did the climate turn cold enough to condense most of it, allowing it to stay mostly at the poles, frozen? Or does it remain as glaciers or permafrost, swaddled under thick surface dust elsewhere? Or buried in ancient craters, as seems so far to be the case? There is a bit of water in the form of ice crystal clouds, too. (There are also dust clouds and clouds made of frozen carbon dioxide—dry ice.) The picture is far from complete on the water issue, and more research information is needed.

> Mars on Earth? Several meteorites from Mars, pieces of the planet itself, have been found on our own planet. They look to be ancient chunks of lava with hardened gas bubbles that match the Martian atmosphere. Long after the Martian volcanic eruptions that created these rocks, a large meteorite must have hit their areas so hard that chunks of the surface were catapulted up into space. Finally the rocks encountered Earth's gravity and were pulled down to us.

The first image of Mars taken by the panoramic camera on the Mars Exploration Rover Spirit. The surface of Mars is reddish in color due to the presence of iron oxide in the soil, much like we see in deserts here on Earth. NASA

Mars, the Life Issue

Even a temporary period of liquid water could have given primitive life as we know it a chance to evolve and to proliferate on Mars. After all, on Earth, organisms live deep under Antarctic ice, in volcanic hot springs, underground in sulfurous caves, and other non-human-friendly environments. Creatures like these are called extremophiles, an ancient life form. There could be Martian versions.

On Mars, methane gas has been discovered in the atmosphere, presumably vented from underground. This is a natural gas associated with swamps. On Earth, some methane is manufactured by living bacteria. Its presence on Mars could indicate some form of primitive, microbial life. Whether we will ever find it is another question.

It is also possible that life once existed on Mars but died out when the liquid water vanished, the planet became cold, and the atmosphere thinned. This did allow dangerous ultraviolet rays from the Sun to reach the Martian surface. If so, fossil evidence of these organisms might someday be found.

Though no one considers it possible to find "little green men or women," here, it is highly possible that very little reddish microbes may be detected, if and when we enter our next round of exploration of Mars.

Quick Tip

1. The existence of several minerals, which form on Earth only in the presence of water, make it likely that Mars had water, though none is visible now.

2. Martian water was probably present in liquid form only for two relatively short periods of its existence as a planet.

Moons of Mars

Mars has just two moons—Phobos and Demos. Both are close to their planet and orbit it in essentially the same plane, in nearly circular orbits. Though that regularity usually indicated that the moons are formed near Mars, the composition of these two objects looks like that of an asteroid.

They may once have indeed been asteroids, existing in the zone of the solar system between the orbits of Mars and Jupiter, then captured by the gravity of Mars. This can happen when Jupiter's powerful gravity disturbs orbits within the asteroid belt and objects assume new orbits.

Moons Beyond/Moons vs. Planets

More moons will surely be discovered in the solar system, especially irregular ones. Their orbits take them far from their planets and at steep angles to the plane that most of the solar system occupies, and this makes them hard to find. These bodies are a quite active area of research in astronomy now, since the ways they were drawn in or catapulted away by their planets can tell us a great deal about conditions in the early solar system. For example, did Jupiter capture so many irregulars because its shape was once much larger when younger—the gases more ballooned out? That could have slowed down some of these objects enough that a capture became possible. In fact, some of the irregulars around Jupiter look as though they were once part of a grouping in the asteroid belt called the Hilda asteroids. Was there a major collision even before their capture? Much more remains to be learned about our smallest solar system inhabitants.

Two other fringe issues relate to the outer edge of the solar system itself. An object found a few years ago, called Sedna, is the most distant body yet discovered in our solar system. This object may be the first from the distant Oort cloud area of our solar system ever to be discovered. Sedna is large—³⁄₄ the diameter of Pluto—and its orbit is irregular. Yet another, called 2003 UB313, or Xena, is both a tiny bit larger and more distant from the Sun than Pluto and even has a moon. Astronomers continue to debate whether Sedna, Xena, a large asteroid named Ceres, and Pluto itself are planets, with the majority now saying that they are not. (See Chapter 7, The Smaller Bodies of Our Solar System, for more on the Kuiper belt and Oort cloud locations in the far outer reaches of our solar system.)

It's a Wrap

In this chapter you have learned that:

 The rocky planets—Mercury, Venus, Earth, and Mars—formed in the inner, denser area of the original solar nebula.

 The moons orbiting both the rocky planets and the gas giant planets are rocky objects; a few are captured asteroids.

 The moons known as "irregulars" are those that do not have orbits close to circular, do not orbit in the same plane as their planet, and do not necessarily even orbit or rotate in the same direction as their planet.

Saturn's moon Titan is the largest moon in the solar system; it has a thick atmosphere, surface pools, and lakes of liquid methane with drier areas in between that include water ice.

Neptune's moon Triton is larger than Pluto.

Meteorites from both the Moon and Mars have landed on the Earth.

 The Earth's Moon formed after a massive rocky object hit the Earth in the early days of the solar system, blasting off the material that later consolidated to make the Moon.

Mercury is not only very small but also has an orbit highly tilted to the plane in which most of the rest of the solar system lies.

Venus has a choking, thick atmosphere, high heat, and volcanism.

Mars almost certainly has water, at least in the form of water ice.

 Though Mars may not have had significant liquid water in many millions of years, it may harbor microbial life.

 More moons are being discovered regularly and they can teach us about the early days of the solar system.

Test Yourself

True/False Questions

1. Jupiter has 12 moons.
2. The moon Titan is coated with organic material.
3. Irregular moons sometimes have originated in the Kuiper belt.
4. A meteorite hit can knock rocks from a planet's space up into space.
5. The Earth's Moon formed about 25 million years ago.
6. Mercury has a thin, hot atmosphere.
7. Venus is cooler on its surface than Mars.
8. Venus rotates on its axis about one hundred times more slowly than does the Earth.
9. The Earth's magnetosphere has made it easier for life to proliferate here.
10. Mars appears reddish because of the high iron content of its soils.

Completion Questions

1. Jupiter has captured some of its moons from _____ _____.
2. Titan, the largest moon in the solar system, belongs to _____.
3. The major moons of Uranus are _____, _____, _____, _____, and _____.
4. The Earth's Moon's diameter is about _____ percent of the Earth's diameter.
5. Venus' surface temperature is about _____ degrees Fahrenheit.
6. Mars has an immense canyon, which slices a line about _____ percent of the way across the whole planet's surface.
7. Mars could have had significant liquid water during _____ periods of its history.
8. If life exists on Mars, it is probably in the form of _____.
9. The moons of Mars are _____ and _____.
10. The two most distant objects discovered in the solar system so far are named _____.

Multiple Choice Questions

1. Jupiter's moon Io is
   a. Home to about 80 volcanoes
   b. Larger than Mars
   c. Made mostly of ice

2. The moon Titan
   a. Shines blue
   b. Has many features formed by water
   c. Was captured from Jupiter

3. Neptune's major moon is
   a. An irregular
   b. A bit smaller than Pluto
   c. One of five moons orbiting Neptune

4. The Earth's Moon is
   a. Tectonically active
   b. Home to frozen seas
   c. Always facing Earth with the same side

5. Mercury's orbit is
   a. The same as the Sun's
   b. 88 days in duration
   c. Distorted by its moon

6. Venus is known for being
   a. Just slightly smaller than Earth
   b. In gravitational control of four moons
   c. A very windy planet

7. Mars has the following feature
   a. Volcanoes that are not considered dead
   b. A thick carbon dioxide atmosphere
   c. Liquid water near its equator

8. Evidence for water on Mars includes
   a. Ancient riverbeds
   b. The mineral olivine
   c. Glaciers discovered under a thick layer of dust

9. The Kuiper belt may contain
   a. More potential moons
   b. Pluto, during the outer part of its orbit
   c. The Oort cloud

10. The moons of the solar system
    a. Look large in the skies of their planets
    b. Are often irregulars
    c. Probably have some underground water

ANSWER KEY

True/False
1. F 2. T 3. T 4. T 5. F 6. F 7. F 8. T 9. T 10. T

Completion Questions
1. The asteroid belt 2. Saturn 3. Titania, Oberon, Umbriel, Ariel, Miranda
4. 25 percent; 5. 870 degrees 6.20 percent; 7. Two 8. Microbes
9. Phobos and Demos 10. Sedna and 2003 UB313 or Xena

Multiple Choice Questions
1. a 2. b 3. a 4. c 5. b 6. a 7. a 8. a 9. a 10. b

CHAPTER 7

◆◆◆◆◆◆◆◆◆◆◆◆◆◆◆◆◆◆◆◆◆◆◆◆◆◆◆

The Smaller Bodies of Our Solar System

◆◆◆◆◆◆◆◆◆◆◆◆◆◆◆◆◆◆◆◆◆◆◆◆◆◆◆

Do I Need to Read This Chapter?

You should read this chapter if you need to review or learn about:

➡ The three main types of smaller bodies

➡ Their age

➡ Location and orbits of asteroids

➡ Role of Jupiter

➡ Composition, size, and shape of asteroids

➡ Difference between asteroids and meteors/meteorites

➡ Types of meteors/meteorites

➡ Types of comets

➡ Location and orbits of comets

➡ Composition, size, and shape of comets

➡ Pluto

➡ Research methods

➡ Smaller bodies studied so far

➡ Meteor showers

➡ Roles in history, in future

Get Started

Three main types of smaller bodies are part of the solar system (and trillions of them exist): asteroids, Kuiper belt objects, and Oort cloud objects. Pluto is probably a Kuiper belt object instead of a planet.

Three Main Types of Smaller Bodies

The three main types of smaller bodies in our solar system—and probably in a goodly number of the universe's other solar systems—are asteroids, Kuiper Belt objects, and Oort cloud objects. Most in the latter two groups become comets if they enter the inner solar system. Meteors and meteorites are not a separate type, as is explained later in the chapter, these are simply asteroids that happen to approach or hit Earth.

Gaspra is a typical asteroid, an irregular rocky body with dimensions about 19 × 12 × 11 kilometers (12 × 7.5 × 7 miles). NASA

None of these bodies is truly large, even the comets with the longest tails. An asteroid's mass, and hence gravity, is low enough that a person could jump right off most of them straight in space. Most of the comets are such loose collections of smaller rocks, dust, and ice that it would be difficult or impossible to stand on them. A comet's loose nature means that its surface ice can vaporize as it approaches the Sun, and its tail is not solid.

These objects are so small that astronomers have discovered only a small fraction of them.

Their Age

Their age, however, is one of the reasons for their importance. Many, if not most, of these asteroids and comets are 4.5–4.6 billion years old. The asteroids, like leftover appetizers from the solar system spread, are bits of the solar nebula that did not typically smash together often enough or hard enough to coalesce into large planets. The material that formed most of the comets and the Kuiper belt objects probably lay originally near where the orbits of Uranus and Neptune are now, then was expelled farther outward within the solar system by the gravity of those bodies. In the outer reaches of the nebula, densities and temperatures were always lower, and it was mostly ice and dirt that clumped together there. Only a couple of relatively massive objects succeeded in forming in this zone, then subsequently suffered enough collisions to split apart. Some of the resulting chunks formed iron-rich bodies and these occasionally reach the Earth as meteorites.

In their outer reaches of the solar system, most of the objects have remained largely unchanged over billions of years, except when they experience more minor collisions or are pulled off course by the gravity of a larger body. Since they usually remain in this cold storage, the compositions and orbits of these objects become keys to the nature of the original nebula around our Sun. Only when the outer objects enter the solar system, as comets, do they warm, releasing material, and thus information.

Quick Tip

These smaller bodies are never older than the solar system itself, 4.5—4.6 billion years old, since they are formed with it.

Location and Orbits of the Asteroids

Most of the asteroids lie in a broad belt between the orbits of Mars and Jupiter. Though about 150,000 of them have been catalogued and named so far, millions more are yet to be discovered. They range in size from pebbles up to 1094 miles (1000 kilometers) in diameter, with most in the low to middle part of this size range. Their neighborhood is large enough that they tend to be spaced at least 1 million kilometers apart.

Although Phobos is a moon of Mars, it did not form around Mars early in the history of the Solar System. Rather, Phobos is an asteroid, captured by the gravity of Mars. NASA

Asteroid Belt Orbits

The asteroids in this location orbit the Sun in the same direction as the planets do. Their orbits, though, are more elliptical and also more highly inclined to the plane of the solar system than those of the planets.

Two Other Kinds

The two other kinds of asteroids, much fewer in number, are the Trojan asteroids and the near-Earth asteroids. The Trojan asteroids belong to Jupiter. One gathering of them orbits at a position about 60 degrees in front of Jupiter, and the other group stays about 60 degrees behind the planet. These asteroids have barely been counted so far, and astronomers expect that there might be hundreds of thousands or more of these small objects. The other kind of asteroid is the group of near-Earth asteroids, whose orbits bring them close to our own. These, which are described later in the chapter, have hit the Earth before and could do so again.

> An unusual asteroid, the consistently closest one to the Sun ever found, was discovered in 2004. Its orbit takes only 6 months. This small addition to our solar system is only about 723 yards—1 mile (500 meters—1 kilometer) in diameter.

Quick Tip

The asteroid belt may seem like a planet that smashed into pieces, but it is actually a planet that never came together.

Role of Jupiter

Jupiter, as the most massive planet in the solar system, has the strongest gravity. Its relative nearness to the asteroid belt gives it power over their orbits. Jupiter acts especially as a shepherd over the Trojan asteroids, nearer to it. But it keeps both these two flocks in place. This is accomplished because of orbital resonance, or orbiting in sync.

Some of the asteroids began originally with orbital periods, half as fast as Jupiter's, and others with orbits in some other simple ratio to Jupiter's (not only one-half and one-fourth but two-fifths and other round numbers work here). Most, actually, have orbital periods somewhere between one-fourth and one-half of Jupiter's.

> Though it may sound unbelievable, a very few asteroids have been found that reached Earth via the Moon and via Mars. Both the Moon and Mars have been hit by many large asteroids in the past, and this kicked up considerable rock debris. Some of it escaped entirely. The orbits of these planetary rocks eventually crossed ours, and they were pulled down to the Earth's surface, to become called meteorites.

Quick Tips

1. The asteroids in the asteroid belt orbit the Sun more quickly than Jupiter's 12-year orbital period because they are closer to it.

2. Since the asteroid belt is so large, and the asteroids so different in mass and in their location vis à vis each other, many somewhat different local motions are possible.

3. The Trojan asteroids' orbits, 60 degrees in front and 60 degrees behind Jupiter, are based on the orbital circle of 360 degrees.

Take as an example an asteroid whose orbital period began as 6 years. Jupiter's, remember, is 12 years. So every 12 years Jupiter catches up with the asteroid in that position—and both are then near each, like two runners on a track when one catches up with the other though their lanes are far apart. When they are close, Jupiter can tug at the asteroid. Over time the asteroid is pulled a bit out of that position (though not into Jupiter).

Because of orbital resonances, certain areas of the asteroid belt actually become temporarily emptied of asteroids. (New asteroids can sometimes arrive later in the gaps, though, since asteroid orbits can be elliptical enough to "slice" through a gap.) Some of the disrupted asteroids may smash into other asteroids—though they are so far apart that this is rare. Over eons, though, the process keeps the asteroid belt an active place.

Some of that activity results in asteroids that are "kicked" by Jupiter into a near-Earth orbit. This possibly unfortunate situation is discussed in the last section of the chapter.

Composition, Size, and Shape of Asteroids

The asteroids are either mostly rocky or mostly metallic. Their masses and densities vary a great deal and are hard to measure from Earth since most are very small and far away. They appear only as a point of light in a telescope or are not even visible at all, though there are several hundred thousand of them.

Size Estimates

The best way to judge the size of a small asteroid with an Earth-based telescope is to compare the light it reflects from the Sun with the light that comes from the thermal radiation of the asteroid itself. The overall brightness of an asteroid comes from its size (bigger is brighter), its distance from us (closer is brighter), and this reflectivity of light from it. The bigger it is, the more it reflects. And the more it reflects, the less radiation remains to heat the object and to be emitted.

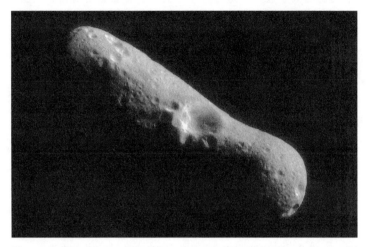

The asteroid Eros was the target of the Near Earth Asteroid Rendezvous (NEAR Shoemaker) mission. Note the large impact crater near the center of Eros. NASA

The surface of Eros from an altitude of 700 meters (2,300 ft.) shortly before the impact of NEAR on the surface. The large, oblong rock casting a big shadow measures 7.4 meters (24 feet) across. NASA

The distance of the asteroid can usually be figured out, first by knowing its orbit and then by comparing its reflected light to its emitted light. With these kinds of information, its size can also be estimated.

Quick Tips

1. There are so many small asteroids because, over several billion years, they have smashed each other into smaller pieces.

2. In other cases, though, the asteroids stick together, making one larger body. This is more rare than a smash-up. The reason is that the asteroid belt is cold, and cold metals and rocks do not meld easily.

3. Asteroids collide when their orbits are changed by Jupiter.

4. They also collide because their elliptical orbits overlap, bringing them up to each other.

5. These collisions are rare, though, since the asteroids of the asteroid belt are so far apart.

6. A collision occurs about once every 100,000 years.

Shape

The asteroids of the asteroid belt are gradually getting smaller and more numerous. Collisions and orbital tugs have also made for many oddly shaped asteroids. Some are like dog bones, many resemble the shape of a russet potato, others are gravelly, others have ridges and troughs, and some even have large craters formed by the impact of a neighboring asteroid which was not strong enough to break it apart. Almost none of the asteroids are truly spherical the way the planets are.

> Asteroid belts surely exist in many other solar systems and one has recently been found around a Sun-like star, about 41 light-years away.

Difference between Asteroids and Meteors/Meteorites

An asteroid can be called a *meteoroid* while it is out in space; later it becomes a meteor when it enters Earth's atmosphere; and it is known as a *meteorite* when it actually hits land or water on the surface of the Earth. Astronomers have learned the most about the larger asteroids by studying the smaller meteorites, even though their fast path through the atmosphere vaporizes a great deal of their mass and leaves their surface burnt and pitted. Almost every meteorite studied so far—about 20,000 of them—has come from the asteroid belt.

Ceres, the largest asteroid, is spherical and about 584 miles (940 kilometers) in diameter, both quite unusual features. Under its thin crust of dirt and carbonate compounds, there may be a thick layer of ice, then a rocky metal core. The ice, if melted, would create more water than all the freshwater on the Earth. Astronomers are now considering Ceres to be a failed planet, prevented only by Jupiter from amassing enough material to be a true planet. Along with Sedna and Xena (described in the previous chapter), and along with Pluto, Certes is in a separate category.

The next two largest asteroids—Pallas (336 miles or 540 kilometers long) and Vesta (317 miles or 510 kilometers long)—are both elongated.

Types of Meteors/Meteorites

In size, meteoroids are small, most are of the size of softballs, pebbles, or even peas. They remain meteors and do not become meteorites even when approaching our planet because they burn out completely before reaching the Earth's surface. The size of the largest meteorites ranges from boulders to massive rocks as large as Manhattan. The latter, very very rare, are the ones that change Earth's history, as is described in the last section of this chapter.

In composition, meteorites can be distinguished from ordinary Earth rocks by the presence of elements no longer found on our surface. Iridium, for example, is found in meteorites, while the Earth's iridium (a rare element) sank to the core of our planet when it solidified long ago.

In origin, meteorites are either "primitive" or "processed." Most are primitive, dating back 4.5–4.6 million years to the birth of our solar system. Of these most are stony, with metallic flakes, and some are carbon-rich, occasionally holding a bit of water. Rarer are the processed meteorites, once part of a larger object which had a core, mantle, and crust, as Earth does, but was smashed apart to contribute to the asteroid belt. Some of these are largely iron/nickel, from the cores of their parent asteroids. Others are less metallic and may even be basaltic, reflecting an origin in the upper layers of the asteroid. (Basalt is the rock that volcanoes belch up.)

Meteorites can conceivably be found anywhere, from a backyard to an ocean bottom, but they are too rare to be discovered often. One exception is in the dry valleys of Antarctica. Here, the glaciers and ice flows of the past have shoved them into specific valleys, and, over the eons, quite a few have collected in these valleys. Nothing has ever disturbed them here. And the fierce winds and other atmospheric patterns in these valleys have kept them from being covered by snow.

Quick Tips

1. Meteorites usually contain elements like iridium, unlike Earth's surface rocks.
2. Primitive meteorites are the age the Earth was when it first formed, 4.5—4.6 billion years old.
3. A smaller number of meteorites, the processed meteorites, were once part of a larger body.

A picture of an iron meteorite laying on the surface of Mars. All of the planets in the Solar System are periodically hit by small debris from interplanetary space. NASA

This meteorite actually came from the planet Mars. It presumably was blasted off into space as the result of an impact on the surface of Mars in the past. It subsequently wandered around the inner Solar System until it struck the Earth in Antarctica. NASA

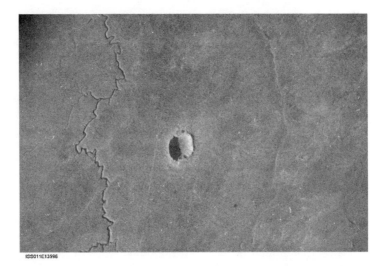

The famous Meteor Crater in Arizona. This crater is less than 100,000 years old, but is already showing signs of the effects of erosion that will eventually erase it from the Earth's surface. NASA

Types of Comets

The other two types of smaller bodies in the solar system are the Kuiper belt objects and the Oort cloud objects. Almost all of them remain in the outer reaches of the solar system, tail-less, though they are usually called comets. Several are discovered every year, when their orbits bring them close enough to us to detect, and especially, as they approach even closer, when their interaction with the Sun's heat and gravity sweeps out a tail. Some comets come too close to the Sun—within 497,097 miles (800,000 kilometers) of its surface—and are partly or completely subliminated. These are called sun-grazers. Other comets return regularly, while some (again because of their orbits) have been here once but never will return again. Other objects in these four outer zones of our solar system are sometimes large enough to be candidates for the label "planet," as discussed later in the chapter.

Hubble Space Telescope image of comet Linear during its dramatic breakup. Comets are fragile objects and often break up into smaller pieces as the orbit the Sun. NASA

It is not a traffic jam out there, but the Kuiper belt and the Oort cloud hold trillions of these objects. Each is a "dirty snowball," or "icy dirt ball" made of ice and rocky bits the size of pebbles down to the size of dust particles. Consistencies range from porous to fluffy. These materials tell us that most of these bodies formed in the outer, colder regions of the solar system, probably about 4.5 billion years right when the solar system formed. (A smaller number joined us in a galactic collision as discussed in Chapter 14, Galaxy Formation and Development.) A few, on their way inward, have been captured by the outer planets, to become moons, but most will never take on that role.

Location and Orbits of Comets (and More)

The two reservoirs of comets, the Kuiper belt and the Oort cloud, are both far away from Earth. Comets are loosened from them by gravitational disruptions from another such object, a planet, or even a passing star.

Segment B of Comet Schwassmann-Wachmann 3, breaking up as it passes by the Earth in April, 2006. This comet broke up into several large pieces on its last approach to the Sun and is ejecting house sized lumps of comet material during this passage. NASA

The Kuiper belt begins at the orbit of Neptune and continues out far beyond the orbit of Pluto, for a range of 30–100 AU, with the best estimates now positing 70 AU as its limit. It contains myriads of the proto-comets and a few of the larger objects, for a total of about 100,000 objects of which only 1000 have been firmly identified so far. About 11 percent of these objects even have moons to accompany them. The potential comets—which range in size from tiny dust balls to more than 621 miles (1000 kilometers) across—are well controlled by the Sun, orbiting in the same direction and often in the same plane as the rest of the planets. Most come closer every couple of hundred to a couple of thousand years, though not always close enough to the Sun to become visible. They are carbon rich, part of the raw material for life as we know it. One, Tempel 1, was recently blasted by a probe from Earth and found to contain clays, carbonates, and crystalline silicates, as well as organic molecules. About 20 percent of these objects even have a moon or an accompanying object. Astronomers are still trying to settle on a good estimate of the total number of Kuiper belt objects.

The far outlying object Xena, mentioned in Chapter 6, The Solar Nebula and Formation of Rocky Bodies and Moons, is a Kuiper belt object. It, along with other very large and newly discovered objects—2005 FY9, also called Buffy, and 2003 EL61 orbit the Sun at such an incline (up to 47 degrees from the plane of the solar system) that astronomers had never thought to look there before.

One outer body, called Sedna, has such an unusual orbit and rotation that it probably came to us either from the Kuiper belt or entirely from another solar system. (More about it is found in Chapter 6, The Solar Nebula and the Formation of the Rocky Planets and Moons.)

The Oort cloud is even more like the "wild west frontier." In this sphere surrounding the solar system on all sides and extending out to about 50,000 AU, all comets have different orbits, leading them in all different directions. Though the cloud is subject to the Sun's gravity, our star's force is slack at these distances, which allows some of the Oort cloud comets to escape regularly as another star tugs at them. Some of them are halfway to other stars to begin with. But the Oort cloud contains about 1 trillion comets, a large enough supply to indicate that comets will be passing relatively near our Sun essentially forever.

Composition, Size, and Shape of Comets

The nicknames "dirty snowball" and "snowy dirtball" (depending upon the balance of dust and ice) comes from the nucleus, or body, of the comet. Before a comet approaches the Sun, this is its only part. Not necessarily round, the nucleus is a lump of dark rocky material and ice.

As a comet is affected by the heat of the Sun, its icy gases begin to sublimate, mixing with some of the rocky bits to form a kind of cloud-like atmosphere of its own. This is called its coma. A coma begins to form when the comet is about as close to the Sun as the orbit of Jupiter or Mars. The coma becomes the bright head of the comet visible to us. As it gets closer, it becomes brighter and brighter as more material is burned off the nucleus. Comets lose so much weight with each pass by the Sun that they eventually vanish entirely, even if they are not sun-grazers.

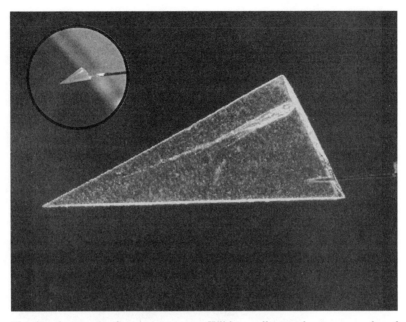

The Stardust probe flew out to comet Wild to collect and return samples of cometary dust. This image shows the track left by a comet particle in a piece of aerogel, a very fluffy, lightweight ceramic, that has been returned to Earth for study. NASA

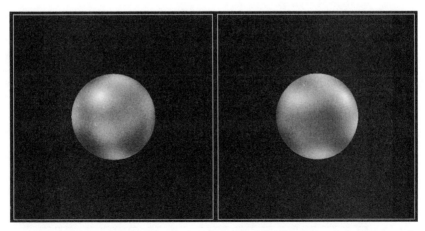

An image of comet Wild at the close approach of the Stardust mission spacecraft. The depressions in the surface may be areas where the comet has lost material to space, or impact craters. NASA

Comet Temple 1 at the moment of impact by an impactor from the Deep Impact probe. The goal was to break a hole in the surface of the comet and observe the dust, ice and gases released into space. NASA

Comet Tails

By the time a comet has reached about the same distance from the Sun as our own orbit, it also sprouts two tails. These can be millions of kilometers long. First one, the plasma tail, is made of the escaping gases—ionized by energy from the Sun. The other, the dust tail, is composed of the rocky bits forced out of the nucleus and coma by the Sun's radiative energy.

Both tails point away from the Sun. And both disappear, along with the coma, after the comet has rounded the Sun and moved back out to approximately the orbit of Mars. There, what is left of the nucleus speeds away.

Comets always shed material in their wake, pebbles and dust that lie along their former orbit until some other object encounters them, as described later in the chapter.

Halley's comet: This most famous comet returns every 76 years. Look for it next in the year 2061.

Gradual Comet Deaths

Every time a comet makes its way into the inner solar system, it shortens its life. This is true whether or not it actually passes very close to the Sun. The average comet in the outer solar system is about 6.2 miles (10 kilometers) in diameter. The Sun "bites off" some, and a comet can lose yet more material if it passes too close to one of the planets. Occasionally, a comet actually crashes into a planet, and this has been documented on both Jupiter and Earth. The most recent Earth encounter occurred in a remote forest of Siberia in 1908.

Quick Tips

1. The three parts of an active comet are its nucleus, coma, and tail.
2. Two types of tail exist: the plasma tail and the dust tail.
3. Sublimation is the process by which solids pass into a gaseous state.

Pluto

Most of us were taught in elementary school that Pluto is the ninth planet. But this body, accompanied by its anomalously large moon Charon and two additional moons recently discovered, are so atypical that astronomers now consider Pluto to be the biggest body in the Kuiper belt and not a planet at all. This is especially plausible when the other large Kuiper belt objects mentioned earlier are considered and compared to it.

Why Different

Pluto resembles the ones in the Kuiper belt instead—in these ways:

- Its orbit is elliptical and, at 248 years, long
- This orbit takes it nearer to the Sun than Neptune for a part of its year
- Its orbit takes it as far away from the Sun as 50 AU
- It is more the size of Kuiper belt object, at 721 miles (1160 kilometers) in diameter, than it is the other planets, smaller even than some of the moons of the solar system
- Its surface is rocky, like Kuiper belt objects, though not as dark

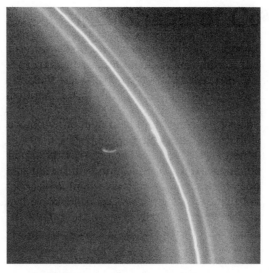

The rings of Saturn are made up of countless, small icy bodies with an typical size comparable to a dining room table. This image shows the sparse F ring and the gravitational influence on the orbits of the ring particles by the nearby small moon Prometheus (the oblong object inside the ring). NASA

No, these are not the rings of Saturn, they are the rings of Uranus. Unlike the rings of Saturn, Uranus' rings are made up of tiny, dark specks of material. NASA

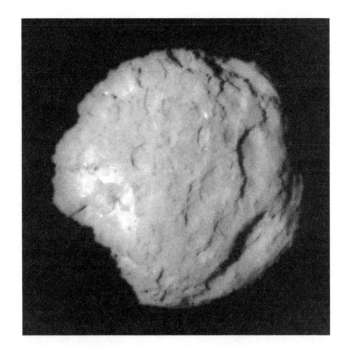

Pluto is very likely a former member of the Kuiper Belt, a large assemblage of icy worlds in orbit around the Sun beyond the orbit of Neptune. It was probably not formed in a manner similar to the gas giants like Jupiter or the terrestrial planets like the Earth. Many of the comets that enter the inner Solar System came from the Kuiper Belt. NASA

- Its rotation axis (north-south pole) is not upright or even nearly so, as is the case with all the other planets except for Uranus—instead the axis is tilted almost to the horizontal position
- It has one very large moon compared to the size of the planet itself and two much smaller ones

Quick Tip

Most of the sky has still not been surveyed to look for other Pluto-like bodies. Astronomers expect there to be others.

Research Methods

The smaller bodies of our solar system are hard to research. Two methods are the easiest. The first is to pick up a meteorite that has fallen to Earth—and about 20,000 of these have been studied, as mentioned earlier. This has enabled scientists to verify asteroid compositions.

A second way is to intercept them up in the atmosphere, and this can be—and has been—done (safely) with the tiniest ones, the meteoric dust. Astronomers have equipped airplane wings with sticky panels to collect these particles. Cosmic dust is always sifting down, since small pebble-like meteors—the "litter" of the solar system—burn up in the upper atmosphere daily. In fact, 100,000 tons of this material falls on Earth every day. Anyone who stays outside for any length of time will go home, sporting some cosmic dust.

Searching

For the smaller bodies that remain in outer space, other research methods exist. Searching by telescope is possible and often succeeds—an asteroid can be spotted because its closer distance to us means that it appears to be moving against the background of stars. Analyzing its light profile or spectrum, mentioned earlier, can then be done in the lab. Well-trained amateur astronomers do most of the discovery work in this area, right from their backyards or parks. They know how to distinguish an asteroid or a comet from an airplane or weather satellite and are allowed to officially name an object they find.

About 60 pairs of asteroids, orbiting each other, have even been found. For asteroids both bright enough and changeable in their brightness, the shape of the body can also be discovered—a spherical body does not change brightness as it rotates, while an elongated one is brighter when its larger size faces us.

In a quite unusual study opportunity, astronomers in 1994 watched a comet smash into Jupiter. By telescope, they observed fireballs, plumes of rising gas that glowed, and massive dust clouds. Jupiter is such a large planet, and composed of permeable gases, that it probably suffered no ill effects.

Visiting asteroids and comets by spacecraft is still rare. But for asteroids that have indeed been orbited by spacecraft, the gravitational interaction between spacecraft and body can be analyzed to learn the density of the asteroid. Smaller bodies with moons of their own also allow for this kind of comparison and knowledge.

Some of Smaller Bodies Studied in Detail So Far

- Gaspra, 9.9 miles (16 kilometers) long, shaped like a russet potato, seen by the Galileo spacecraft en route to Jupiter
- Mathilde, 36.7 miles (59 kilometers) long, shaped like a rock punched by another rock and very low in density, seen by the Near-Earth Asteroid Rendezvous spacecraft on its way to Eros (see below)

- Eros, 24.9 miles (40 kilometers) long, almost the density of solid rock, orbited and then landed upon by the above-named spacecraft

- Ida, 32. 9 miles (53 kilometers) long, shaped like a mummy in a museum, seen by Galileo en route to Jupiter. Ida has a tiny moon all of its own.

- Comet Wild was approached by a spacecraft in 2004, with the spacecraft on its way back to Earth with material from the comet. On site, the NASA craft saw features of the comet from craters, mesas, and spires to jets of gas and dust spouting off the surface. Also detected were the nitrogen-carbon bonds needed to launch life in hospitable places, such as on the early Earth, when it, too, was visited by comets like this one. Wild's consistency (though not its composition) seems to be like freeze-dried ice cream. Look for more results in 2007 and 2008.

Meteor Showers

Meteors are the opposite of rare—about 25 million hit the top of the Earth's atmosphere every day. These tiny bits of rock, traveling at about 155,343 miles (250,000 kilometers) per hour, vaporize completely from the heat generated by their friction with the Earth's atmosphere. Most burn out quite high, at about 31–62 miles (50–100 kilometers) above the Earth's surface.

Meteor showers are meteoroids that flame out as a group event. The grouping is created by a broad path of pebbly and dusty debris left by a comet on its orbit through the solar system. These debris paths, like cosmic gravel trails, remain in place for long periods of time because the comets' elliptical orbits do not overlap with the Earth's completely. So we swing through only part of a given comet's broad trash path every year. That intercepts cometary material.

The Perseid meteor shower, for example, occurs when we cross the debris path of the comet Swift-Tuttle, every year. (Look for it at its height on August 12 every year, the birthday of this book's first author.) The comet itself no longer exists.

Four Top Meteor Showers and the Dates on Which the Maximum Number Can Be Seen

Quadrantids: January 3
Perseids: August 12
Taurids: November 3
Leonids: November 17

Because of meteor collisions and gravitational interactions with larger bodies (which sometimes can eject the smaller bodies from the solar system), there are fewer smaller bodies around our solar system now than there were in earlier days, with the exception of the asteroid belt. Then, meteorites pelted the Earth regularly.

Roles in History, in Future

So far, in human history, a couple of meteorites have smashed into the backs of cars, one has seriously injured a dog, and one did crash through a roof to bounce off a table and bruise a woman's leg. But meteorites have not killed anyone. Of course, accurate historical records go back only a thousand years or so, and the existence of people on Earth who could even keep records goes back just a few thousand years. Until the last few hundred years, too, there were so few of us on the planet that potentially dangerous meteorite hits could have easily happened with no one nearby to get hurt or even to notice.

Large meteorite hits have indeed happened in the distant past, however. One probably caused the extinction of the dinosaurs about 65 million years ago and possibly other major extinctions in Earth's history. The space rock does not need to hit creatures on the head. The tremendous bomb of a major hit kicks up so much surface material (from rock to water) into the atmosphere that it blocks much of the Sun's light and heat. This debris spreads through the whole atmosphere as the Earth rotates, darkening everything. As the months and then the years drag on, first the plants die, then the smaller creatures that eat the plants die. Next the larger creatures that eat the smaller creatures can no longer find any food. And of course a giant hole is punched into the ground with tremendous force, killing everything in its path.

Even if the meteorite hits the ocean, the scenario is still drastic. Massive tsunamis kill many land creatures immediately.

Asteroid 2004 MN$_4$ will provide quite a sight in 2029 when it approaches close enough to Earth to be seen by the naked eye. Though it will be faint since it is so small—350 yards (320 meters) across—this asteroid will approach us as close as 18,600 miles (29,934 kilometers) from the Earth's surface. It is considered quite safe. But, in astronomical terms, this is a near miss.

Early Days

In the earlier days of the solar system—the first 2 billion years—there were many more asteroids out there and all the planets were pummeled by them regularly. Some astronomers think, too, that early hits by comets and meteorites may have brought our planet its first water and first organic molecules, delivered straight from outer space.

Few people have seen evidence of these massive meteorite hits of the past, but that is only because they are hard to see unless you know what to look for. Craters created by meteorites are much easier to see on the Moon. On Earth, crater edges collapse, mountain ranges rise, islands are created, glaciers spread thick blankets of dirt—all these forces conceal the "evidence of the crime." And, of course, some of the impact craters are on the bottom of the ocean.

Scientists have indeed found, though, about 170 meteorite craters on Earth, so far. These range from 0.5 miles (0.8 kilometers) to about 285 miles (400 kilometers) across. Some, since they were ancient hits, are now under layers of soil and rock. These have been spotted because rocks that have been stressed by the original impact, called tektites, breccias, and shocked quartz, can sometimes still be lying on the surface. A few craters are actually easily visible, such as the giant hole punched into the desert in northern Arizona. Called Meteor Crater, this one is a tourist attraction.

Will an immense meteorite pound our planet again? Yes. The near-Earth asteroids, mentioned earlier, could hit, since their orbits cross ours (though at elliptical angles). Astronomers are now plotting the orbits of all those known to exist. Studies have, however, already missed some (these, fortunately, whizzed by).

Every asteroid plotted receives a number from 1 to 10 based upon its perceived risk. This "Torino scale" can be found at http://www.neo.jpl.nasa.gov/torino_scale.html (and listings of nearby asteroids are available at http://www.spaceweather.com). The project, called Spaceguard Survey, is managed by the National Aeronautic and Space Association (NASA). What to do if a major risk presents itself is being researched; one project is looking at a "Tugboat" effect or a "Tractorbeam" project, in which the problem asteroid is deflected without being touched in any way.

Another "near miss" of an asteroid encounter occurred in spring of 2005. Coming within a zone four times the distance of the Moon from the Earth, 4179 Toutatis passed us by. It is a potato-shaped rock, about 3.1 miles (5 kilometers) by 1.2 miles (2 kilometers). Gone for now.

It's a Wrap

In this chapter you have learned that:

 The three main types of smaller bodies in the solar system are asteroids, Kuiper belt objects, and Oort cloud objects.

 Most of these smaller bodies are 4.5–4.6 billion years old, dating to the beginning of the solar system.

✔ The types of asteroids are asteroid belt asteroids, Trojan asteroids, and near-Earth asteroids.

✔ Jupiter affects the orbits of many asteroids.

✔ A meteor is an asteroid (or, very rarely, a comet) that has entered the Earth's atmosphere.

✔ A meteorite is an asteroid that hits the Earth's surface.

✔ Comets are "dirty snowballs" or "snowy dirtballs."

✔ Most comets remain in the outer solar system forever, though some enter the inner solar system and some escape to orbit other stars.

✔ Pluto is probably a Kuiper belt object.

✔ Meteorite impacts have changed Earth's history and could do so again.

Test Yourself

True/False Questions

1. The smaller bodies of the solar system were once a large planet that exploded.
2. About 150,000 asteroids lie in the asteroid belt.
3. Some asteroids have been found that were originally part of the surface of Mars.

4. Jupiter's gravity often pulls asteroids into its atmosphere.
5. Asteroids and comets are rarely spherical.
6. A rock found on Earth with a blackened crust might be an asteroid.
7. Sun-grazer comets are the ones that crash into the Sun.
8. The Kuiper belt has fewer objects than the Oort cloud.
9. Comets have two tails when they are moving through the inner solar system.
10. Pluto is 1160 kilometers in diameter.

Completion Questions

1. The asteroids and comets of the solar system are about _____ years old.

2. The asteroid belt lies between the orbits of _____ and _____.

3. A collision occurs between asteroids in the asteroid belt about every _____ years.

4. The greatest natural collection of meteorites found on Earth is in _____.

5. The Oort cloud extends as far out as _____ AU from the Sun.

6. The three parts of an active comet are _____, _____, and _____.

7. Asteroids and comets studied by spacecraft are _____ in number so far.

8. Meteor showers happen when the Earth intersects _____.

9. The time when the most meteorites fell on the Earth was the first _____ years of the solar system's existence.

10. The type of asteroid most likely to hit the Earth is called a _____ asteroid.

Multiple Choice Questions

1. Which of the following is not a type of smaller body in the solar system?
   a. Trojan asteroid
   b. Kuiper belt object
   c. Mercury

2. What is the most likely shape of an asteroid?
   a. Russet potato
   b. Sphere
   c. Disk

3. How fast do meteors travel?
   a. 250,000 kilometers per hour
   b. 885 kilometers per hour
   c. 80 kilometers per hour

4. Why are large meteorite impact sites hard to find?
   a. The craters fill up and becomes lakes.
   b. The craters become eroded.
   c. The impact killed all the people in the area.

ANSWER KEY

True/False
1. F   2. F   3. T   4. F   5. T   6. T   7. F   8. T   9. T   10. T

Completion Questions
1. 4.5–4.6 billion years   2. Mars and Jupiter   3. 100,000   4. Antarctica
5. 50,000   6. Nucleus, coma, tail   7. 5   8. The path of an old comet
9. 2 billion   10. Near-Earth

Multiple Choice
1. c   2. a   3. a   4. b

CHAPTER 8

The Sun's Role in Our Solar System

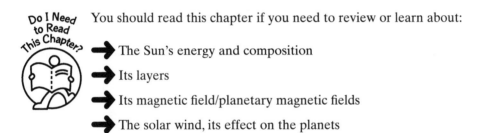

Do I Need to Read This Chapter? You should read this chapter if you need to review or learn about:

➡ The Sun's energy and composition

➡ Its layers

➡ Its magnetic field/planetary magnetic fields

➡ The solar wind, its effect on the planets

Get Started

Though some people may think that the Sun's role on Earth is in the emission of its light and heat, that is far from complete. Our planet, and all the planets, are bathed in its solar wind every day.

Sun's Energy and Composition

Though stars are the subject of later chapters in the book (Chapter 12, Star Formation, Star Development and Chapter 13, The Deaths of Stars), our local star—the Sun—is so important in this region of space that its basic nature and effects upon our solar system are described here. Its luminous energy is what makes life possible on Earth, and that energy is prodigious. Only 1 second's worth of it, if entirely captured on this planet, would fulfill the world's energy needs for half a million years. The Sun's life as a sphere of gaseous matter, called plasma, began about 4.6 billion years ago at the center of a solar nebula. (For more information, see Chapter 5, The Solar Nebula and the Formation of Gas Planets.) Our star will burn, nuclear fusion raging in its core, for about 5–6 billion years more.

In the meantime, the Sun's energetic output includes not only visible light but also ultraviolet light, x-rays, infrared light, radio waves, and gamma rays. Our atmosphere protects us from much of this radiation, which is truly fortunate, since the ultraviolet light would vastly increase the incidence of cancer and the gamma rays and x-rays would probably kill us all before that.

The Sun is composed of fiery hot gases in the form of a plasma. It has only a small core where the gas is so dense that it is considered solid.

Quick Tip

The Sun is a mid-life star, about halfway through its natural lifespan. Its death will end life as we know it on Earth.

A picture of the active Sun in ultraviolet showing a very large prominence rising above the surface. NASA

The Sun's Statistics

- Diameter: 864,947 miles (1,392,000 kilometers)
- Temperature at core: 15 million degrees Kelvin
- Constituents: 70 percent hydrogen, 28 percent helium, 2 percent heavier elements
- Temperature at surface: 5800 degrees Kelvin, cooler within the sunspot storms

---

Sun's Layers

Farthest away from its core, and considered as features rather than layers, the Sun frequently emits both solar flares and coronal mass ejections. Intermittent, energetic events like these accelerate streams of protons and electrons out from a small specific region of the Sun's surface—perhaps only as large as 20 times the entire surface area of the Earth.

The solar flares, gaseous but resembling giant licks of flame, can shoot out very far. In 2004, for example, an especially large spate of them was detected by the

Cassini spacecraft when the craft was farther than 808 million miles (1.3 billion kilometers) from the Sun; the Voyager 2 space probe felt flare effects at 6.6 billion miles (10.6 billion kilometers) out. Flares may even race out even farther than this. They are bursts of x-rays that originate near sunspots, described below.

The coronal mass ejections are like giant blobs of magnetized gases, weighing about a billion tons apiece, occasionally much more. They belch out from the Sun. One, detected in November of 2004, was the fastest and hottest ever recorded—it ballooned out at 5 million miles (8 million kilometers) per hour and at 41 million degrees Celsius. It missed the Earth only because it blasted out from the side of the Sun not facing us at that time. This was fortunate since its energy could have caused a nearly worldwide electrical blackout.

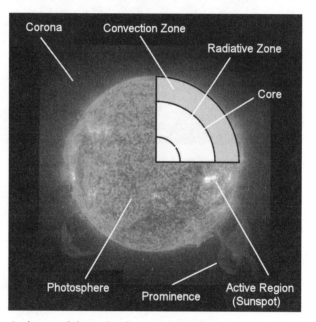

A picture of the active Sun overlaid with a cutaway showing the layers of the Sun. TJJ

These ejection events certainly boost the intensity of the everyday solar wind reaching us, too, as is discussed later in this chapter.

Corona

As a kind of crown or nimbus around the Sun, this outer level of the Sun's true atmosphere is thin. It is also extraordinarily hot at about 1 million degrees Kelvin. This is the part of the Sun seen in photographs of a solar eclipse.

Sunspots

These magnetic storms on the outer surface of the Sun tend to be like small dark "potholes"—perhaps only the size of a few Earths put together. Their temperatures average about 3500 degrees Celsius, which is cooler than the rest of the Sun's surface. (This is what makes them appear to be black.) Sunspots seem to move around the Sun, but this is only because the Sun itself rotates on its axis. Sometimes there are just one or two weaker sunspots visible, sometimes there may be a dozen large ones. They wax and wane in an 11-year cycle, from what is called the solar maximum to the solar minimum. These highly electromagnetized gases explode out, adding plenty of extra material to the ordinary solar wind. Effects on Earth are discussed at the end of the chapter.

Other stars seem to have sunspot cycles also. Two stars where solar flares have been detected are Beta Persei and V711 Tauri, both similar to the Sun when it was a bit younger in its life as a star.

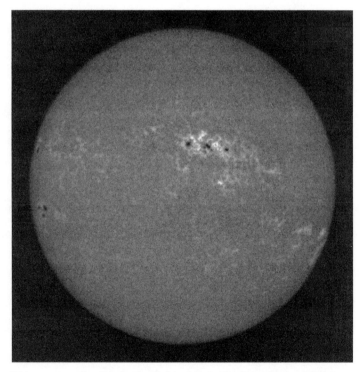

Here we see the Sun in the light of emission from ionized Calcium. Notice the sunspots that show up as dark patches with bright emission on their periphery. NASA

Nearer to Surface

The chromosphere and the photosphere, dotted with shifting rough granules about 186 miles (300 kilometers) high, and the surface itself, with giant valleys and peaks of scorching gas, are other features of our neighborhood star. The chromosphere blazes at about 4000–10000 degrees Kelvin. The photosphere, at about 6000 degrees Kelvin, looks like the actual surface but is still a quite thin gas. It is the region of the Sun that emits the visible light we see from Earth. Sunspots are primarily on the photosphere.

In the Sun's core operates the basic force of nature called the strong force. Existing at the subatomic level and only over these small distances, it overcomes the electromagnetic force (which makes positive charges repel each other) to bind together positively charged protons and neutrons as they collide at high speeds in the fiery gaseous heat. As the hydrogen nuclei fuse into the helium nuclei, binding tight, the result is nuclear fusion. (See Chapter 4, Key Concepts and Basic Laws, for an introduction to basic forces of nature.)

An image of the Sun in the extreme ultraviolet showing activity rising from the Chromosphere up into the inner Corona. NASA

The Depths

Below the Sun's photosphere is the convection zone where gases at different temperatures seethe and roil. Under this lies the radiative zone, where temperatures of about 10 million degrees Kelvin excite photons and electrons.

The core is the true fusion furnace, with temperatures of about 15 million degrees Kelvin. Here the pressure is about 200 billion times greater than that on the Earth's surface. This temperature and pressure allows the Sun to provide us with the energy required for life on Earth.

Quick Tip

The Sun's nuclear fusion creates energy which takes about 1 million years to rise up through its outer layers—then about 8 minutes to move from its surface to Earth, at the speed of light (of course).

Sun's Magnetic Field/Planetary Magnetic Fields

At the immense energies that exist on the Sun, magnetic fields are created as the charged particles in the plasma travel along specific magnetic lines of force. Looking a bit like lines of iron fillings around a small magnet, they wrap around the Sun and direct energy outward from the surface. The magnetic energy pulses a bit stronger in the sunspots because magnetic fields can trap extra pockets of gas. These conditions are always changing as small differences of temperature and pressure cause turbulence within the plasma. Thus the sunspots form, dissolve, then reform elsewhere. Holes in the corona allow the charged particles to sweep out of the Sun entirely. This creates the solar wind, which is described in the next section.

The magnetosphere, the term for the nimbus of magnetism surrounding the Sun, is not only a feature of stars. Planets, too, can have magnetospheres. Their magnetism emerges from hot iron and other metals in their cores (a different mechanism from the one that obtains on the Sun).

Magnetic field lines anchored in the surface of the Sun guide
hot gas along large arcs up through the Chromosphere and
into the Corona. NASA

A computer generated image of the hot tubes of gas in the
inner Corona of the Sun. These tubes are confined by the
magnetic field of the Sun. NASA

Planetary Versions

Deep within a rocky planet's core and mantle, convection from heat and pressure can cause molten metals to seethe. This material then also moves as the planet spins on its axis. These motions can then throw off an invisible nimbus of magnetism, strongest along the magnetic lines of force surrounding the planet. These planetary magnetospheres look like a miniature version of the Sun's, though the Sun has no metallic core.

The planets in our solar system which have magnetospheres are: Mercury, Earth, Mars, Jupiter, Saturn, Uranus, and Neptune. These magnetospheres are of vastly differing strengths. Among the rocky inner planets, the Earth's is the strongest, Mars the weakest. But each magnetosphere works to protect its planet from the forms of solar radiation, dangerous to life as we know it, acting as a kind of shield.

Quick Tip

Planets with magnetospheres are those which have cores still hot with molten metal, convection within the core, and sufficiently fast spin on the planetary axis to keep everything moving.

Solar Wind and Its Effects

As described earlier in the chapter, the Sun emits a solar wind of charged particles. The wind's speed averages about 310 miles (500 kilometers) per second. It moves out in all directions to the edge of the solar system. Its extent is, in fact, one definition of that edge, the end of the sway of the Sun and the beginning of interstellar space.

Sunspots boost the ordinary amount of protons and electrons sweeping out and thus the strength of the solar wind. This material reaches the Earth's atmosphere and enters along our magnetic lines of force. After a sunspot "storm," it usually takes 24–48 hours for the energetic particles to hit the Earth's atmosphere. Coronal mass ejections often arrive sooner, in about 12 hours. The website www.spaceweather.com can thus tell when Northern or Southern Lights are likely to appear to the observers under the right conditions and in the right positions.

If a chronograph is used to block the intense light from the photosphere of the Sun, the fainter, but much hotter Corona can be seen. This is similar to what takes place during a total eclipse of the Sun. National Center for Atmospheric Research (NCAR)

Results

When billions of tons of the Sun hit the small bubble of atmosphere surrounding our tiny planet, it is not surprising that effects is noticeable. The solar wind compresses the Earth's magnetosphere, and particles at the top of the Earth's atmosphere are accelerated in bursts of energy, electrons are stripped from atoms of the various gases there. The energy races down along the magnetic lines of force of the Earth's magnetosphere, stretching them like a slingshot—ready to fire. Our shield becomes breached, in these places, by our active star itself.

Some 60–240 miles (96.5–386 kilometers) above sea level, the fireworks can begin. This phenomenon is called the Northern Lights, or Aurora Borealis, and their near mirror image in the Southern Hemisphere is the Southern Lights, or Aurora Australis. The energy can easily be a billion trillion watts.

This auroral energy via the solar wind can also be quite disruptive: satellites can be disabled, power lines damaged, radio communications broken up, and airplane passengers bathed in some extra radiation. These problems occur during the most powerful displays.

For those who live or visit at latitudes of about 45–85 degrees north in the Northern Hemisphere or south in the Southern Hemisphere, aurora are much more common since the magnetic lines of force are closer together at the North and South Poles. It is the experience of a lifetime to see a good display of Northern Lights in dark, clear, moonless skies. They may begin in the north as leaping blurs of light, then arc up like pale flames. Sometimes they look like they are thick white light, folded, or curtains of green pulsating, or a red blur spilled across half the sky, or arcs like milk pouring, beginning at the zenith and cascading down the sky. Every display is different, since the energy and the amount of matter of their origin are different. Some of the Sun is appearing at night, in a sense, right in our backyards. (Those living closer to the equator, at latitudes such as that of New Mexico, for example, may be treated to this astounding phenomenon only once every few years, after an especially strong storm on the Sun.)

Quick Tip

Just as the Sun's sunspot cycle involves 11 years of strengthening to a peak and then 11 years of weakening, so does the frequency and impressiveness of Northern Lights displays. But even supposedly "low" years can be amazing. Some of the most dazzling displays of the last few years have occurred in the weakening phase.

Other planets in our solar system are also known to have aurora. So far, they have been detected on Jupiter, Saturn, and Mars.

In this chapter, you have learned that:

 The Sun emits radiation across the whole electromagnetic spectrum: gamma rays, x-rays, ultraviolet light, visible light, infrared light, and radio waves.

✔ The Sun has layers within its plasma, with its core the home of nuclear fusion, in which hydrogen and helium nuclei fuse

✔ The Sun is about halfway through its lifespan

✔ The solar wind, enhanced by solar flares, sunspot storms, and coronal mass ejections, reaches the Earth constantly and sometimes creates aurora displays there

✔ The Sun and most of the planets have magnetospheres

Test Yourself

True/False Questions

1. The Sun's temperature increases steadily from its corona to its core.
2. The Sun creates Northern Lights, or aurora, every day.
3. Magnetic lines of force are visible around the Sun but not around the Earth.
4. Magnetism from the Sun helps to operate magnets on Earth.

Completion Questions

1. The major layers of the Sun, beginning from its center are _____, _____, _____, _____, _____, _____.

2. The strong force, as it operates on the Sun, results in the fusion of nuclei of _____ with _____.

3. An average coronal mass ejection weighs about _____.

4. Energy from the Sun's nuclear fission takes about _____ to rise up through its outer layers.

Multiple Choice Questions

1. The solar wind sweeps out as far as
   a. The Earth's orbit
   b. Neptune's orbit
   c. The edge of the solar system

2. Nuclear fusion in the Sun
   a. Will last about 5 billion years longer
   b. Occurs at about 15 million degrees Kelvin
   c. Both of the above

3. The Sun is composed of
   a. Mostly hydrogen and helium
   b. Magnetism from metals
   c. Hot, solid gases

4. After a sunspot storm, the aurora can be seen in clear skies at high latitudes
   a. 1–2 days later
   b. Everywhere from Mars to Pluto
   c. As a red blur

ANSWER KEY

True/False Questions

1. F   2. F   3. F   4. F

Completion Questions

1. Core, radiation zone, convection zone, photosphere, chromosphere, and corona 2. Hydrogen/helium 3. 1 billion tons 4. 1 million years

Multiple Choice Questions

1. c 2. c 3. a 4. a

PART III

Astrophysics Basics

CHAPTER 9

Origins of the Puzzles of Space-time and Gravity

Do I Need to Read This Chapter?

You should read this chapter if you need to review or learn about:

➡ The first era of modern astronomy

➡ Einstein's two theories of relativity

➡ How anyone with a home computer can help extend his theories today

Get Started

Though Einstein's theories on space-time are well accepted, tests of them are still being conducted. And gravity has not yet been completely integrated into current theory.

Launch of Modern Astronomy

Though people have probably always been curious about the beginning of the universe in space and time and about how it is held together, the modern investigation of space-time and gravity began only in the 1500s. That was when the astronomer Nicholas Copernicus first argued for a Sun-centered—heliocentric—solar system, though with the planets in circular orbits, not their elliptical orbits, and when Tycho Brahe gathered the first accurate astronomical data, setting up what could be called the original "big science" laboratory. Brahe's elaborate observation center on the island of Ven, off Denmark (now Sweden), was complete with staff, visiting scientists, and assistants who functioned much like today's graduate

Tycho Brahe was the last great astronomer to use his naked eye for observing. Soon afterward the invention of the telescope revolutionized astronomy. Public Domain

students. Over three decades, Tycho Brahe's group—working even before the invention of the telescope—collected enough precise naked-eye observations of comets, a supernova, and much more to further push the heliocentric theory. (Up until this era, people had believed that the Sun orbited the Earth. After all, they saw it rise, move across the sky, and set every day on the other side.)

By the early 1600s, Johannes Kepler used Tycho's observations (especially of the orbit of Mars) and formulated theory to state three laws of planetary motion. The first posited that each planet moves around the Sun in an elliptical orbit. The second demonstrated that planets travel faster when nearer to the Sun than they do in outer parts of their orbits. And the third stated that planets whose orbits lie farther from the Sun move more slowly than do the inner planets. He came very close to describing gravity, in this case the Sun's pull on the planets.

This stamp portrays Johannes Kepler, who made use Tycho Brahe's observations to discover that the orbit of Mars was an ellipse, not a circle. Public Domain

It took Isaac Newton, in the late 1600s, to explain the gravitational force of the Sun on solar system bodies. He accomplished a tour de force of mathematics that still describes the large-scale motions of the solar system accurately. It took more than 250 years for additional complexities to be uncovered.

Galileo Galilei, who used the recently invented telescope in the early 1600s, was also a giant of early astronomy. His mathematical studies of the mechanics of astronomical bodies, how they slow and accelerate, for example, were as important as his discoveries of solar system objects such as the four largest moons of Jupiter.

Portrait of Galileo Galilei by Giusto Sustermans. Galileo was the first scientist to use mathematics and perform experiments in a manner similar to modern scientists. Public Domain

The cover of Galileo's book "Dialog Concerning Two Chief Systems of the World" (1632) portrays three individuals. One represents the ideas of Galileo, one a rational skeptic that becomes convinced by Galileo's arguments and one who never abandons the old idea of an Earth centered universe. Public Domain

---

Quick Tip

1. Nicholas Copernicus, Tycho Brahe, Johannes Kepler, Galileo Galilei, and Isaac Newton are the chief figures in the beginning era of modern astronomy.

2. Many actual observations, as well as theories, were required.

The Einstein Revolution: Special Theory of Relativity

In the first of his two theories of relativity, Albert Einstein confronted and explained what occurs in the universe at speeds beyond the ordinary scale of the solar system and its Newtonian motions; namely, that space and time are related to each other mathematically. His special theory of relativity, formulated based on the contributions of many scientists and published in the very early 1900s, presented this concept of space-time. In it, the speed of light is a constant, never changing, and an observer at any point in the universe would see the forces of nature as equivalent.

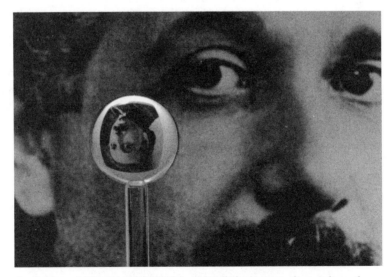

A picture of Einstein reflected in one of the extremely precise spheres used in Gravity Probe B. NASA, Gravity Probe B

Space and time are linked in these ways:

- An object moving at close to the speed of light (186,000 miles or 299,338 kilometers per second) appears to an observer to be experiencing a time slowdown. A clock on that object would be behind when compared to the time on the observer's own wristwatch.

- An observer seeing two lights far out in space in different directions but flashing at the same time would not agree with a second observer atop an object moving close to the speed of light; the second observer would not see the two flashes as simultaneous.

- An observer watching an object that was streaking by at near to the speed of light would see the object as shrunken in the direction it was moving, almost as though it were compressed into a pancake.

- The streaking object would also appear to be more massive. The "m" is for mass, in $E = mc^2$.

- No object with a finite *rest* mass can reach the speed of light.

The reason that the theory is called *relativity* is that is posits that the motion of an object is relative to the position of the observer in space and time. Where you are and when you look governs what you see.

This theory is highly relevant for astronomy since the universe is laced together with many, very fast-moving, massive bodies. For a listing of the different speeds at which even our modest planet, the Earth, is traveling, see Chapter 2, The Earth's Place. (These numbers are stated from the perspective of an observer above the surface of the Earth. Observers in different positions would see us differently.)

Though Einstein's special theory of relativity may appear a bit dizzying, it is anchored in two solid features: the speed of light is the same everywhere in space-time, and the laws of nature (such as this theory) are consistent across the universe. As will appear in the rest of the chapter and in the two following chapters, Einstein's theories, as far as they go, have been corroborated by a century of experimental evidence.

The Einstein Revolution: General Theory of Relativity

In his second theory, Einstein attempted to make gravity part of his mathematical explanatory system and worked primarily on his own to do so. At scales beyond our everyday life on this planet, the gravitational force of massive objects such as stars and entire groupings of galaxies holds the entire universe together. (It also operates on Earth, of course, making apples fall down, not up, from trees and keeping the atmosphere we breathe from escaping into space, among other helpful phenomena.)

Gravity may seem to be a kind of invisible taffy, stretched between two bodies to hold them together in space. But it is not so. Gravity is actually what happens when space-time is distorted. The mass of astronomical bodies pulls at the "fabric" of space-time, and these distortions are what govern the motions of the bodies, pulling them into their differing courses. Stronger gravitational forces can actually slow down time considerably. The effects of gravity on a body are the same as the effects of acceleration. And supermassive objects in space that accelerate in some way actually emit gravitational waves. Astronomers are still seeking evidence of these waves, posited by complex theory.

These gravitational waves move at the speed of light. (Objects that have mass cannot do this, but massless particles and waves such as light itself can.) The nature of gravitational waves is part of the next chapter, which takes Einstein's, and later theories, down to the atomic and subatomic levels. Both the macro and the micro levels are important in astronomy since conditions inside stars, for example, are stripped down to the most basic physics and chemistry.

Gravity Probe B was able to measure an effect due to the rotation of the Earth on the local space time continuum called "frame dragging," predicted by Einstein's General Theory of Relativity, by using the most precise gyroscopes ever made. NASA

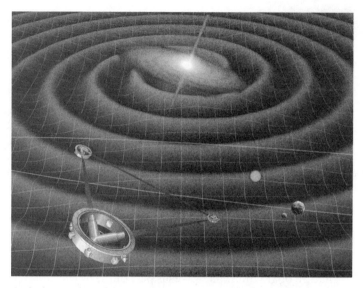

Artist impression of the gravity waves emitted as two black holes spiral inward together. The proposed NASA probe to measure gravitational waves from space, LISA, is illustrated in the foreground. NASA-JPL

Quick Tip

The expression "time is the fourth dimension" is phrased imperfectly. Time is just one of the four dimensions, the other three being length, width, and depth. Time is merely the only one of them we cannot see.

Helping Einstein at home is actually possible. On your PC (Windows, Linux, or Mac), begin by setting up an account at http://einstein.phys.uwm.edu. The program installs very quickly.

Then, anytime your computer is idle, it can help search for gravitational wave signals from supermassive neutron stars, which have been focused upon for this purpose at general observatories. The data, from detectors in Washington, Louisiana, and Germany, have been gathered from neutron stars up to 1500 light-years away. Finding direct evidence of gravitational waves emitted by them would help integrate the forces of nature itself, as described further in the next chapter.

General relativity is highly significant for an object approaching a black hole. This actually happens regularly across the universe, and results in stars, clouds of interstellar gas, and the like being "swallowed" by the gravitational pull of the black hole. The immense gravity of a black hole distorts space-time, bending light until it cannot escape from the black hole. (For more on black holes, see Chapter 13, The Deaths of Stars.)

It's a Wrap

In this chapter you have learned that:

✔ Copernicus, Brahe, Kepler, Galileo, and Newton provided the foundations for modern astronomy.

✔ Einstein's Special Theory of Relativity shows how space and time are mathematically related.

✔ The speed of light is a constant, and the laws of nature are consistent across the universe.

✔ Einstein's General Theory of Relativity posits the effects of gravity upon space-time.

✔ Gravity waves have not yet been actually observed, but an effort is on to do so.

Test Yourself

True/False Questions

1. Galileo set up the first large-scale science lab.
2. Newton's laws of planetary motion have been proved wrong by Einstein's theories of relativity.
3. We may someday be able to travel close to but not at the speed of light.
4. In essence, the theory of relativity establishes that everything is relative.
5. The gravity of the Earth and other small astronomical bodies also distort light and space-time, but the effects are too small to be observable or important.

Completion Questions

1. Kepler's first law of planetary motion states that _____.

2. The speed of light is _____ per second.

3. Gravity waves move at the speed _____.

Multiple Choice Questions

1. To an observer watching another object move faster and faster toward the speed of light, that object would
   a. Slow down
   b. Increase in length and mass
   c. Both

2. The effects of gravity
   a. Are explained in Einstein's Special Theory of Relativity
   b. Are the same as the effects of acceleration
   c. Indicate that the universe will eventually contract

ANSWER KEY

True/False
1. F   2. F   3. T   4. F   5. T

Completion Questions
1. Each planet moves around the Sun in an elliptical orbit   2. 186,000 miles
3. Of light

Multiple Choice Questions
1. c   2. b

CHAPTER 10

Down to the Atomic and Subatomic Levels

Do I Need to Read This Chapter? You should read this chapter if you need to review or learn about:

➡ The four forces of nature

➡ The elements of matter and energy

➡ Quantum mechanics

➡ Quantum states

➡ The uncertainty principle

Get Started

Though scientists regularly make progress on this issue, we still do not know how the four most basic forces of nature are interrelated.

Four Basic Forces of Nature

There seem to be basic four forces, which together describe the universe as we know it:

- *The strong force.* Also known as the strong nuclear force, it holds protons and neutrons together inside the nucleus of an atom. This force, without which we would have no solid objects (including no people), operates only over very small distances, at least in the universe as we know it today. The fusion that occurs in a star such as our Sun involves the strong force.

- *The weak force.* Also known as the weak nuclear force, it controls some nuclear reactions, such as the production of radioactive Beta decay and of neutrinos. Only weak in comparison to the strong force, it also operates only at infinitesimal distances in the universe as we know it today.

- *Electromagnetism.* This force, an amalgamation of electricity and magnetism, holds whole atoms together (and everything made of them) and produces the radiation emitted by astronomical objects. It operates at a large-scale compared to the two above forces, and can be quite strong.

- *Gravity.* Though this force is observably strong, it is actually both the weakest (about 10^{37} time weaker—10 times10, 37 times—than the strong force) and the least understood. It holds together stars, planets, and galaxies, but how it relates to the other three forces is perhaps the major puzzle of physics today.

Unifying the Four Forces

Scientists have developed a Grand Unified Theory (GUT) in which the first three forces are linked mathematically, seen as aspects of the same force. They are separated now, the theory goes, only because the universe has cooled since its birth. Though the first three have thus been unified, the fourth force, gravity, has still not been incorporated into the theory. The goal of that effort is called the search for the "Theory of Everything."

Is it possible that gravity was joined to the first three forces at the immense temperatures that existed only at the very birth of the universe? Scientists think so, although more research and more theory needs to be done on this question. For example, no one has yet observed a graviton, the elementary particle that carries this gravitational force, conveying its energy. Gluons are the particles that carry the strong force, weak bosons carry the weak force, and photons carry the electromagnetic force (including light). All of these elementary particles are better understood than the graviton.

Elements of Matter and Energy

The smallest particles of nature, all subatomic in that they are parts of an atom, include, as of now, neutrons, protons, and electrons, supplemented by positrons, gluons, photons, and many others. These particles vary in electrical charge, in spin, and in which of the two groups they belong to—fermions or bosons. Not just tiny blobs, they also behave as though they were waves. They are both matter and energy, though we call them particles.

Beneath the level of the particles mentioned above, and even smaller, since they are within the neutrons and protons, are the leptons and the quarks. There are six types of each. The quarks are usually, though not always, found as a threesome inside an atom. Along with these, there are always representatives of the antimatter world, which appear and disappear, such as the antileptons and antiquarks. (The antimatter world is described in Chapter 17, The Future of the Universe, where its relevance to the ultimate fate of the universe makes it more pertinent).

Quick Tips

1. Elementary particles are both matter (objects) and waves (energy).
2. The most basic of these, the building blocks, are the six leptons and the six quarks.

Though work with neutrons, protons, and electrons may seem to be relevant only to theory and to laboratory experiments, this is not the case. The Big Bang that began the universe was at one time composed entirely of energy, a force field, only forming the first elementary particles later. Billions of years later, the neutron stars, of which the universe contains many, are still made of this early material: 90 percent neutrons, with the rest protons and electrons. (These unusual stars are discussed in Chapter 13, The Deaths of Stars.) And it is a subatomic effect that prevents the cores of dying stars from collapsing entirely. These are only three examples of the real world relevance of subatomic study.

Do even smaller particles exist in the subatomic world? Are there unknown processes or even more forces of nature? If there are, that could mean space has dimensions beyond those we can perceive, or even conceive of. The theory of this possibility is known as *supersymmetry*. How the heaviest quarks disintegrate and whether quarks come in groups beyond three and five are among the lines of research which should prove fruitful in answering these questions ultimately. It is possible that the four forces of nature and the array of elementary particles which obey them can all be explained in terms of a theory called *superstring* theory. In this theory, the universe is made of tiny strings which vibrate in supersymmetry across 10 or 11 dimensions. This area of astrophysics is proving to be more and more interesting and provocative. Some of its puzzles are addressed in Chapter 17, The Future of the Universe.

Quantum Mechanics

This is the term used to describe the theory of how the elementary particles of the atom, from the larger proton to the tinier quark, interact with other elementary particles at various energy levels. The field of quantum mechanics can be thought of as the third major stage of astronomy, following the birth of modern astronomy and the achievements of relativity theory.

As of now, six types of quarks are believed to exist—labeled with the curious names of "up, down, strange, bottom, and top." Unlike the larger objects of our world, quarks can flicker in and out of existence in microseconds, even in virtually empty outer space, given a bit of energy to work with. Using the mathematics of quantum mechanics, scientists can discern the probability, but not the certainty, that a particle will be at a certain place, in a certain time, and at a certain energy level. This is called the uncertainty principle, which is described later in the chapter. They also have learned that no two fermions (e.g., protons, electrons, and neutrons) can be in the same quantum state at the same time. This is called the exclusion principle. Quantum effects exist at the edges of black holes and are highly relevant in other ways to the evolution of the universe.

Quantum States

One of the reasons that subatomic physics is so tricky to study is that elementary particles, using immense amounts of energy, may change their spin, their position, their magnetic charge, and their whole energy state (their quantum states) often and in microseconds of time. A particle's quantum state is a description of its location, momentum, angular momentum (as it orbits the atom's nucleus), and its spin at a particular moment. Since these particles make up astronomical objects such as stars, gas clouds, and black holes, their behavior is highly relevant for astronomy; yet re-creating it for analysis in a laboratory on Earth requires devices such as particle accelerators which are as large as several football fields and cost millions of dollars. (For more about these accelerators, see Chapter 18, Frontiers in Astronomy.)

Uncertainty Principle

Because of the rapid changes in the universe's elementary particles and the fact that they are both matter and energy, these constituents of the universe are more like smears or blurs than solid objects. They pop in and out of existence and, even when "alive," cannot be precisely located. This situation creates uncertainty at the most basic level of the atom (and of the universe, which is made of atoms). The theory that describes it is called the *uncertainty principle*.

The uncertainty principle can be stated in various ways. One way is that the more we know about a particle's location, the less we know about its mass and movement (its momentum) and the more we know about its momentum, the less we know about where it is; another way is to say that the instruments we use to detect a particle, since they involve light, magnetism, or some other "tool," actually change the energy state of the particle. Its dimensions and behavior cannot be pinned down just because you are trying to find it. There is thus randomness—probability rather than certainty—at the heart of matter and energy, though it exists at such an infinitesimal level that we do not experience it directly.

Quick Tip

Two of the fundamental principles of quantum mechanics are the uncertainty principle and the exclusion principle.

Quantum teleportation seems to be something straight out of science fiction. Could it ever really be possible to separate a person into packets of quanta and transport them on laser beams for reassembly into the person at the destination? "Beam me up, Scotty" (from the old Star Trek) may never be safe for humans, but versions of it using laser pulses have indeed succeeded in appearing to move an ion of one atom over to another atom. The two atoms correlate energy states without touching each other or interacting in any other way.

The most recent quantum teleportation method developed in a laboratory has used electronic fields near to absolute zero in temperature to line up ions, then blasted them with a careful sequence of laser pulses to alter the energy state of the first ion and then transfer that energy state to the other ions. They seemed to jump from one place to another.

It's a Wrap

In this chapter you have learned that:

 The four basic forces of nature are the strong force, the weak force, electromagnetism, and gravity.

 Elementary particles include electrons, neutrons, protons, which make up atoms, and the particles which carry the forces, such as the gluon.

At the most basic level lie the six leptons and six quarks (as well as their antimatter counterparts.)

Subatomic theory operates at the tiniest level and also throughout the large scale of the universe

Quantum mechanics, following the evolution of modern astronomy and Einstein's relativity theory, describes the way particles and energy interact

Test Yourself

True/False Questions
1. Gravity is strong than the strong force.
2. There are usually three quarks in every atom.
3. Antimatter is science fiction, rather than science.
4. Particles can change their quantum states.

Completion Questions
1. The two forces that operate at the subatomic level are the _____ and _____.
2. Scientists think that the forces of nature and the elementary particles controlled by them may ultimately be seen as one theory called _____.
3. The six types of quarks are _____, _____, _____, _____, _____, and _____.
4. The particle that carries the gravitational force is called the _____.

Multiple Choice Questions

1. Quantum teleportation indicates that
   a. Objects can be split up into their subatomic particles and transmitted elsewhere for reassembly.
   b. The first two forces of nature have been unified.
   c. Gravity has not yet been unified into a GUT.

2. Quarks can
   a. Change their spin
   b. Flicker in and out of existence
   c. Both

ANSWER KEY

True/False

1. F   2. T   3. F   4. T

Completion Questions

1. The strong forces and the weak force   2. Supersymmetry or superstrings
3. Up, down, strange, bottom, and top   4. Graviton

Multiple Choice Questions

1. a   2. c

CHAPTER 11

How These Are Studied

Do I Need to Read This Chapter? You should read this chapter if you need to review or learn about:

➜ What methods are used to test subatomic theories from Einstein's relativity to quantum mechanics

➜ Particle accelerators and quantum mechanics

➜ Laser studies and space gases

➜ Radio telescope research and black holes

➜ Satellite-based studies

Methods Used

In addition to the human brain (the most impressive analytic device of all) and a supercomputer to analyze massive amounts of data according to mathematical formulas, scientists have devices to verify, and extend, aspects of theories such as Einstein's relativity and quantum mechanics. These include particle accelerators, lasers, telescopes such as radio telescopes, and satellite-based studies. (Additional devices used to study astronomical phenomena are described in Chapter 18, Frontiers in Astronomy).

Perhaps the most exotic research method is the use of a massive particle accelerator—an enclosed ring, often miles in diameter, with room for whole golf courses in the middle. Only a few of these accelerators exist around the world, and they are used by international teams of scientists.

In a particle accelerator, subatomic particles are accelerated as close as possible to the speed of light (186,000 miles or 299,338 kilometers per second). As they race around inside the tube, like invisible horses in a Kentucky Derby, they often collide. In such energetic events, some of the energy turns into mass, following the Einstein equation $E = mc^2$ ("E" is energy, "m" is mass, and "c" is the speed of light). This mass is other, new elementary particles, sometimes those that would never be observable unless a scientist were present at the Big Bang or dived inside a star. These particles live for only nanoseconds. But that can be long enough to change theories of cosmology.

Particle acceleration demonstrates something else, too, every time—no object with a *rest mass* can reach the speed of light. No matter how much energy accelerates the particles in these massive machines, not one of them has ever reached that speed.

The muon g-2 storage ring at Brookhaven National Laboratories. This ring was used to make the first precise measurement of how negatively charged muons "wobble" in the magnetic field, information which can be used to confirm the Standard Model of particle physics. Courtesy of Brookhaven National Laboratory

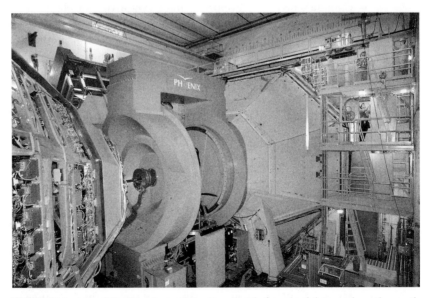

PHENIX is one of the four large detectors that helps physicists analyze the particle collisions at Brookhaven's Relativistic Heavy Ion Collider (RHIC). Courtesy of Brookhaven National Laboratory

Particle Accelerators and Quantum Mechanics

Of the numerous experiments conducted with particle accelerators, just six will be mentioned as recent examples:

- Brookhaven National Laboratory's "Relativistic Heavy Ion Collider," in New York state, seems to have created the substance formed in the very infancy of the universe, before matter existed at all: the quark-gluon plasma. (This substance is described more fully in Chapter 1, The Universe in Time and Space.) A fiery gas, trillions of degrees hot, this plasma contains loose quarks and gluons.

- German's HERA Collider, having created millions of collisions over 4 years, seems to have found about 50 "pentaquarks" (five quark particles, each containing the hard-to-find "charm quark"), a new type of elementary particle.

- Japan's particle accelerator, the KEK, has detected a new form of decayed "bottom quark" and its antimatter companion (before they annihilated each other, which was almost instantly). One of the four forces of nature, the strong force, governs this atomic decay. This particular form of decay indicates that additional elementary particles, never discovered so far, might even exist and be affecting the process.

- Canada's TRIUMF cyclotron and a cyclotron in Indiana detected a rare type of imbalance of asymmetry (called a charge symmetry violation), which explains the balance between protons and neutrons in the universe. The same experiment may lead to estimates of the mass of the up and the down quark, which has not yet been determined, and even to the discovery of the "strings" of "string theory," mentioned in the previous chapter and to be mentioned in the book's last chapter (Chapter 10, Down to the Atomic and Subatomic Levels and Chapter 18, Frontiers in Astronomy).

- Europe's laboratory on the Swiss-French border, with its new Large Hadron Collider, will attempt to accelerate particles to such high energy (though not to the speed of light) that strings and even tiny black holes may conceivably appear. Accelerator-created particles vanish quickly because of all the energy in the system and because of quantum mechanical flux.

- A sensing system built to extend c. 450 miles (724 kilometers), underground, from the Chicago area to northern Minnesota, is studying of neutrinos, tiny subatomic particles with little or no mass and no electrical charge. These mysterious particles pass right through rock, and our bodies, constantly. They also swish right through the planet, which includes the rock between Chicago and Minnesota. These neutrinos are so common that whether or not they have mass significantly affects the weight of the universe and hence its future. (Please see, Chapter 18, Frontiers in Astronomy for more detail.)

A remarkable experiment dubbed Ice Cube will measure elusive cosmic neutrinos by drilling dozens of holes over 2 km deep in the Antarctic ice at the South Pole. In this picture, sensitive light detectors are being lowered into the holes to view rare neutrino interactions through the clear, solid ice under the surface of the Antarctic. National Science Foundation

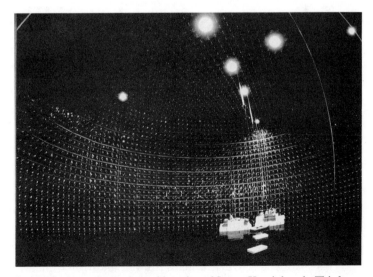

Inside the large dome shaped interior of Super-Kamiokande. This large volume is surrounded by over 11,000 light detectors and is filled with 50,000 tons of pure water when in operation. Like Ice Cube, Super-K is searching for cosmic neutrinos. Courtesy Kamioka Observatory, ICRR (Institute for Cosmic Ray Research), The University of Tokyo

Quick Tip

Particle accelerators, incredibly complex and expensive machines, are not numerous, but they are our best key for testing old theories and developing new ones about the nature of the subatomic world and of the universe.

Laser Studies and Space Gases

Lasers of all kinds are used a great deal in subatomic research. One example of this was the use of lasers at temperatures near to absolute zero in a high vacuum chamber. (The chamber was chilled and literally vacuumed out of particles irrelevant to the experiment.) Tiny clouds of lithium gas were manipulated by the lasers, which then were turned off. Instead of moving outward to fill the chamber, the gas formed itself into a pancake. And at these temperatures, it behaved as though it were a single particle. This kind of unusual expansion, asymmetric, is thought to be the way the early quark-gluon plasma expanded in the first microseconds of the Big Bang at the launch of the universe.

Radio Telescope Research and Black Holes

Radio telescopes—those which detect radio waves instead of other forms of radioactivity such as visible light—have been used to find the shadows of black holes. This is the zone where light from nearby stars or gas clouds is being bent by the black hole. At a mass of nearly 4 million Suns, the black hole at the center of the Milky Way, for example, musters powerful gravity. Space-time itself is warped by it.

Throughout most of the other chapters of the book, the focus is on the findings, with no room available to discuss how each was made. This chapter and Chapter 18, Frontiers in Astronomy, are designed as glimpses behind the scenes and hints at the process of "doing science."

The shape of the black hole's shadow is also relevant because, if the black hole is rotating (as it probably is), it will be pulling space-time around with it. This phenomenon is called frame-dragging.

This shadow discovery, which required an array of nearly a dozen radio telescopes around the world, is as close as we will ever get to seeing a black hole. And it is the first time it has actually been done.

Black holes can never be seen directly. Their gravity is so strong that light cannot escape them—so there is nothing to see, except for the shadow.

The Keck Observatory telescope with its 10 meter (33 ft.) diameter primary mirror is a good example of modern, large ground based telescopes. The larger the mirror the greater the light gathering power of a telescope. Courtesy W. M. Keck Observatory

The two 10 meter Keck Telescopes are located at the submit of Mauna Kea, a 4200 meter (13,800 ft.) volcanic mountain in Hawaii. Astronomers want to locate their telescopes at high, dry sites to reduce the adverse effects of the Earth's atmosphere as much as possible. Courtesy W. M. Keck Observatory

Satellite Studies

Frame-dragging, mentioned in the previous section, is also important because it may contribute to the formation of quasars. These are the jets of energy and matter that lie at the center of some galaxies, and they may well be the most powerful objects in the universe. The torque of the warp of space-time around a black hole might manufacture some of (manufactures) this power. And it should be detectable in its small, local level by the gyroscopes on board a NASA satellite. The issue is how much these sensitive devices are thrown off balance by the force of the Earth's own gravity as it slightly both warps and drags space-time.

The satellite will soon report back—is Einstein's theory of relativity precisely correct here? Or is the universe even stranger yet?

Yet another satellite, this one launched by Japan's space agency, carries the coldest scientific instrument ever sent into space. At only six-hundredths of a degree above absolute zero, it will be able to detect the heat of a single x-ray photon. The goal is to study the x-ray emissions of supernovas (exploding stars), the x-rays released as black holes eat gas clouds, and other events that are very energetic but occur many light-years away.

These are just two examples among the many satellite-based studies recently conducted, being conducted right now, and planned for the future.

Attached to the "robot arm" the Hubble Space Telescope is unberthed and lifted up into the sunlight during this the second servicing mission. By flying up above the Earth, the HST can take very sharp images without the blurring effects of the atmosphere. NASA

The Spitzer Space Telescope is the last of NASA's Great Observatories. The SST is specially built to operate at very cold temperatures and observe the universe in the infrared. Here we see the Delta II launch vehicle with the SST on top. NASA

CAD rendering of the Wilkinson Microwave Anisotropy Probe (WMAP) satellite. WMAP has made the most accurate measurements of the Cosmic Microwave Background to date. NASA

Chandra X-Ray Observatory being readied for integration into the launch craft. NASA

 In this chapter you have learned that:

 Particle accelerators, by smashing elementary particles together, have helped to verify Einstein's theories of relativity.

These colliders have also led to insights into the quark-gluon plasma presented shortly after the Big Bang, quarks of various types, asymmetry in the universe, string theory, and other core phenomena that operate at both the subatomic and universe-wide levels.

Lasers, specialized telescopes, and satellite-based instruments have also extended our knowledge of the subatomic world.

Test Yourself

True/False Questions

1. Particle accelerators are probably the best tool for studying the subatomic phenomena relevant to astronomy.
2. These accelerators seem to have created tiny black holes and antimatter.
3. Neutrinos can pass through rock.

Completion Questions

1. The speed of light is _____.
2. In a collider/particle accelerator, elementary particles created typically last for _____.

Multiple Choice Questions

1. Particle accelerator experiments have led to the probable discovery of
   a. Strings
   b. New stars
   c. Planets outside of the solar system
2. Laser research has been used to make
   a. Telescope improvements
   b. Manipulations of gases
   c. Medical lithium

ANSWER KEY

True/False

1. T   2. T   3. T

Completion Questions

1. 186,000 miles per second   2. Nanoseconds

Multiple Choice Questions

1. a   2. b

PART IV

Stars and Galaxies

CHAPTER 12

Star Formation, Star Development

Do I Need to Read This Chapter?

You should read this chapter if you need to review or learn about:

➡ Star birth in the early universe and today

➡ Sizes of new stars

➡ How stars ignite

➡ Star evolution, star types/the H-R diagram

➡ Open cluster, globular clusters

➡ Variable stars

➡ Stars in small groups

➡ Star layers

Star Birth in the Early Universe and Today

The universe just before the birth of the stars was seething with energy and matter. By about 400,000 years after the Big Bang, electrons and protons had first been able to bind together to make neutral hydrogen atoms within the plasma that seethed through the early universe and is still the hottest substance of all anywhere. The electron and proton particles have been considered subatomic, bound into some kind of atom, ever since. All this activity set the stage for stars, but none had yet formed. This time, called the Cosmic Dark Ages, lasted for a few hundreds of millions of years, ending only as the first generation of stars began to be born. They were to be extremely massive ones, up to 150 times the mass of our Sun. (For more on the early universe, see Chapter 1, The Universe in Time and Space.)

These about-to-be stars began to form inside lumpy clouds of molecular gas. The clouds contained only hydrogen and helium, light elements that had formed early in the process of particles being bound into atoms. Hydrogen and helium are the main constituents of the universe then, and now. Temperatures were moderate within these clouds (a few hundred degrees Kelvin). Gravity gradually pulled the material into larger and larger clumps around the original lumpy areas within these diffuse, cool clouds. Each clump became a node for one, a few, or occasionally many, new stars.

As gravity continued to do its work, that work released heat. Enough of this dissipated to keep the clouds relatively cool still. This was important because it allowed the heat pressure outward in the clumps to be counteracted by stronger gravity pressure inward. The cloud began to glow slightly from this energy within it. By the time the gravitational attraction of each clump had strengthened as it grew, most gasses from the cloud spiraled into these nodes. At this stage, each object became what is known as a *protostar*.

As more of this in-falling gas gathered, it concentrated in the center more and more tightly. Eventually the temperature and density at the protostar's core

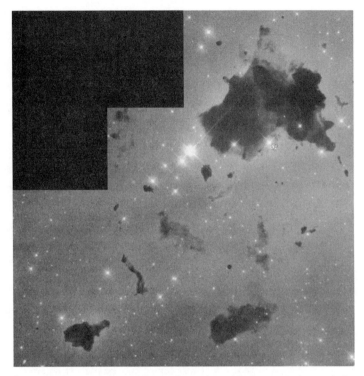

Stars in the current epoch are born in cold, dense regions of the interstellar medium. Here we see dense globules that contain so much dust that they block the light from stars and nebulae shining in the background. NASA

became great enough for nuclear fusion to occur—hydrogen burned into helium and later into heavier elements. Light shone.

These first, hot stars were so massive that they lived very short lives, ending in immense explosions we call *supernovas*. The star explosions blew the heavier elements, such as carbon, oxygen, and iron, out into the surrounding space, seeding space with the material for the next generation of stars.

Stars, however, are still being born regularly. In late 2004, astronomers found a tiny "baby" star, weighing less than one-thousandth of the mass of the Sun. Its name: L1014. Will it accumulate enough gas and dust to burst into full light? Only time will tell—a great deal of time.

When a hot, luminous star forms in a region of gas and dust, it often erodes nearby dark clouds, exposing dense lumps that were inside. These lumps may be the site where the next generation of stars will form. NASA

Today

Though star formation was far more common in the early universe than it is today, it is still occurring. Astronomers can peer into dusty clouds and disks to see protostars emerging even now. They are watching quite a few of star nurseries at this time, some of them in unusual areas, such as in the streamers of gas thrown off when two galaxies collide and others in the hydrogen-gas penumbra of single galaxies well beyond that galaxy's visible limits. Some stars in these areas are very young, massive, hot objects with limited life spans ahead, including several whose surface temperature is higher than 10,000 degrees Kelvin. They will live for probably no longer than 100 million years. Most star formation, though, occurs in the cores of the galaxies (where the surrounding brightness makes it very difficult to observe). (For more on the various regions of galaxies, see Chapter 14, Galaxy Formation and Development, and Chapter 15, The Role of Dark Matter.)

Astronomers have also recently found what looks to be one of the oldest stars in the universe. Formed many billions of years ago, it is probably one of the universe's second generation of stars. Astronomers have labeled it HE 1327-2326.

The Cone Nebula is a 'pillar' of dense gas and dust with star formation taking place near the tip. Buried down inside the upper end of the pillar is a very young star, invisible in this image, shining mostly in the infrared. NASA

A rare find, this faint star lies about 4700 light-years away, within the Milky Way, and it is about the mass of our Sun. At this nearby distance, we are seeing the star as about what it would look like to an up-close observer, too. (Its light left the star only 4700 years ago.) HE 1327-2326 is known to be so old because of its chemical composition. It has only about 1/250,000 as much iron as the Sun does, the lowest amount of iron of any star yet discovered. Iron—a heavy metal and more complex than the hydrogen and helium matter of the early universe—had not yet had time to form in any great amount by the era of this star's birth, since atoms beyond these basics form inside of stars and are ejected back into the universe as raw material for new stars and other astronomical objects when they die. So its low iron content dates it as very old. (Please see Chapter 13, The Deaths of Stars, for more on the elements forged by stars.)

This ancient star also has small amounts of carbon, oxygen, and strontium, typical of a second generation star, since at that time material had already begun to diversify beyond the era of the first-generation stars. Its birth probably occurred just a few million years after the first generation of stars in the universe. Astronomers are seeking more such primitives, even in surprising areas, such as the outskirts of galaxies where leftover gas clouds may still be close in composition to those of the early universe.

Most of the stars and nebulosity in this image would be invisible to the human eye, even with a very powerful telescope. This image of the star forming region DR-21 was made using the Spitzer Space Telescope to peer through the intervening dust at infrared wavelengths. NASA

Quick Tips

1. The first-generation stars were massive, up to 150 times as massive as the Sun.

2. Within cloud nurseries, gravity and heat forces worked in balance to form protostars.

3. Protostars then shed material, emitting winds and jets, which pushed away their nursery "swaddling."

Time Travel to Early Universe

Though it may seem impossible, astronomers can actually see back to the time in the universe when the first generation of stars formed, about 13.7 billion years ago. This is because very distant objects, whose light can take billions of years to reach us (even traveling at the speed of light), are spotted now that the light is finally reaching us.

Time travel like this has allowed astronomers to detect a hot, bright, star-crowded area, where about 1 million, infant first-generation stars have been studied. It is 12 billion light-years away. Since the universe is about 13.7 billion years old, that is a look now, but back to the time when the universe was only about 1.7 billion years old. Other such nurseries have been found, too. This is especially fortunate since the first stars were so massive that their lives were much shorter than the lives of most stars today. The stars in this particular ancient nursery are about

50,000 degrees Kelvin and range from about 10 to 20 times the mass of the Sun (so they are probably among the smallest of their generation).

Most other nurseries where stars were even larger and even shorter-lived have probably vanished entirely. But an even older glimpse has yielded some evidence of them. A faint blur has indeed been detected, all that can be seen of these far, far distant very first-generation stars lying about 13 billion light-years away.

Size of New Stars

Astronomers typically divide stars into "low-mass stars," "high-mass stars," and the "close binaries" mentioned earlier in the chapter. The reason for this is that the evolution and the death of a star depend crucially upon its initial size and also on its position vis à vis other stars.

This division by mass is rough in terms of the universe as a whole, with astronomers using the Sun as a reference point; it is considered to be about average, low-to-middle mass. At this level of mass, stars can last a long time. The Sun happens to be in the middle of its lifespan, also. At not quite 5 billion years

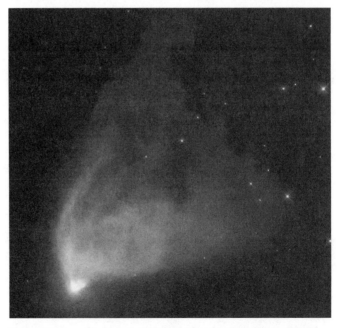

Before stars settles down to become a normal, hydrogen burning main sequence star, they go through considerable instability and activity. Here we see R Mon in the lower left, which is lighting up Hubble's Variable Nebula. This nebula is material thrown off by R Mon over the course of the last few thousand years. NASA

old, our star has about another 5 billion years to go. (For more information on the Sun, see Chapter 8, The Sun's Role in Our Solar System.)

Low-Mass Stars

The most common kind of star in the universe, low-mass stars range from about (half) 1/10th the mass of our Sun to about twice its mass. These stars live the longest, which is one reason that there are so many of them around to observe, about 50 times as many as the most massive of the high-mass stars.

Steadily, these stars fuse hydrogen and helium within their cores. The energy created there can take a million years to rise from the core and flame out the surface as light and other radiation. The process happens by radiative diffusion near the core (as photons bounce between electrons) and by convection (rising and sinking plumes) nearer to the surface. Deeper convection zones with the plumes are typical of low-mass stars. What happens to them several billions of years later is described in Chapter 13, The Deaths of Stars.

High-Mass Stars

These stars, less stable, hotter, and more short-lived, are the primary "engines" that have created and are creating many of the chemical elements of the universe and of the life within it. The upper limit for mass in a star seems to be about 150 times the mass of the Sun. (Beyond that, gravity would have given way to internal pressure and the star would have blown itself apart very quickly after formation.) The first stars were mostly in the very high mass category, unlike today, when most stars are smaller and more like our Sun.

Fusion of hydrogen and helium is faster at the temperatures inside high-mass stars. And the small amounts (c. 2%) of carbon, nitrogen, and oxygen get involved to make it even faster, in a process known as the CNO cycle. With this furious rate of energy generation, the stars radiate light at a prodigious rate. They also create radiative pressure that expels an especially large amount of star material via a stellar wind. This quicker loss of mass is one reason for the short lifetime of these stars—just few million years. More about their death is described in the next chapter.

Quick Tips

1. Low-mass stars radiate most of their energy more (through steady convection) *slowly*, which prolongs their lives.

2. High-mass stars experience tremendous radiation pressure near and at their surface, which strips them of material.

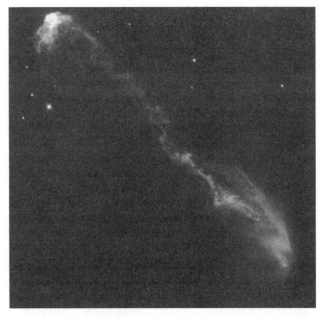

Sometimes the material thrown off by young stars can form astonishingly long jets that travel out from the star and smash into gas in the interstellar medium. This image shows HH-47 (for Herbig-Haro object number 47), a young stellar outflow 3,000 times longer than the distance between the Earth and the Sun. NASA

An image of a debris disk around the red dwarf star AU Mic. Planets may be forming, or might already exist, within it. Astronomers now consider the formation of planetary disks as a universal part of the overall star formation process. NASA

Close Binaries

The universe contains myriad binary stars, which evolve according to their own masses and hence do not form a single separate category based upon mass. The stars known as close binaries lead different lives. They are always born together and so are of the same age. And, being close together, they pull on each other until their shapes become oval. This is just the beginning of their strangeness.

Since one of the stars in a close binary system is usually more massive than the other, the larger one begins to move through its life stages faster (as larger stars do). The more massive one eventually enters the stage that involves swelling, whereupon the smaller one is able to capture some of its companion's outer material since it is now nearer. Eventually, the lower-mass star, whose "health" is better, has consumed so much of its companion that it itself becomes the more massive one. Then its lifespan also begins to shorten. For what happens when two stars like these approach death, see Chapter 13, The Deaths of Stars.

Quick Tips

The three main categories of stars, for the purpose of understanding their evolution, are high-mass stars, low-mass stars, and close binaries.

How Stars Ignite

To move from the end of the protostar stage discussed earlier to become a full-fledged star of any mass, ignition of the star must take place. The energy that produces light, and all the other forms of radiation which stars release, is from nuclear fusion. It begins when the core of the protostar contracts and thus becomes much hotter, even though some of the heat is still radiating away from its surface. By the time the core reaches 10 million degrees Kelvin, hydrogen fusion begins. Hydrogen nuclei fuse unto helium nuclei, releasing tremendous, and sustainable, energy—sustainable for much of the life of the star. (This is why a star cannot truly be said to "burn," though astronomers use the word for convenience: it is not "on fire.")

High-mass stars take no longer than a million years to thus ignite, very low-mass ones more than a hundred million years, and our (average) Sun took about 50 million years to move from the protostar stage to becoming a true star. At this point, the evolution of the life of a star is described in a graph called the H-R diagram.

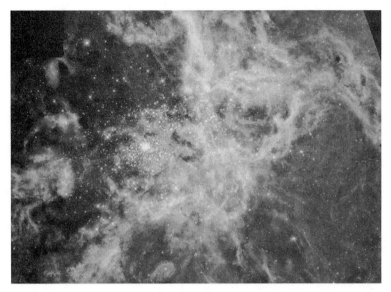

When stars much more massive than the Sun are born, they are so hot and luminous they can strongly alter their local environment. 30 Doradus is a region in the Large Magellanic Cloud that contains several dozen of the most massive stars known (just left of center). Their strong ultraviolet light and fast stellar winds have ionized and sculpted a large volume of surrounding space. NASA

Low Mass Companion to Star Gliese 105A   HST · WFPC2
PRC95-33 · ST ScI OPO · September 14, 1995 · D. Golimowski (JHU), NASA

Gliese 105 is a low mass, faint, red dwarf star not too far from the Sun. In this remarkable image we see a companion to GL 105 (in the upper right) that is not a true star like the Sun. It will not burn hydrogen in its core for an extended period of time. Astronomers call these objects Brown Dwarfs. NASA

Star Evolution, Star Types/The H-R Diagram

Stars can be described in a variety of ways: by their mass, diameter, rotation, surface temperature, chemical composition, radial velocity, and luminosity, for example. The most convenient parameters to use to ponder their evolution, their life stages, are their mass, temperature, and luminosity, those being highly related. (Big stars are usually hotter and brighter, though the exceptions are very interesting, and are mentioned later, too.)

This relationship forms the basis of perhaps the most useful graph in astronomy. It is called the H-R diagram after the two astronomers who separately discovered the key relationship early in the last century, Ejnar Hertzsprung and Henry Norris Russell. There are many variants of the diagram—whatever an astronomer research team is studying about stars can be plotted on its own variant of the diagram.

Main Sequence Stars

The diagram's main band, the arc in the middle, is where virtually every new, fully-fledged star is positioned after it settles down from its birth (and before it moves toward death). Stars typically spend 90 percent of their lives here. And it is where 90 percent of the stars identified in the universe are found today, including our Sun. It is called the main sequence.

Not every star is typical, however. Here are four examples, all recently discovered by astronomers, but clearly with their own styles:

- The smallest star yet known, barely larger in diameter than the planet Jupiter. It orbits a much larger companion, in a regular binary system, and this is what led to its discovery. The little star dimmed the light of the larger star very slightly, as it orbited its large companion, establishing its smaller size. But it is clearly massive enough to exert a gravitational tug on its "big brother."

The flattest star ever found, at least 1.5 times wider than it is tall. This star rotates so fast—at 140 miles (225 kilometers) per second—that it has "squashed" itself.

- The fastest star ever found, speeding fast enough to move the distance from New York City to Cleveland in 1 second. Astronomers think that it may have had a binary companion, and the pair of them approached too close to the black hole at the center of the Milky Way. As its companion fell in, this star was propelled out. The speedster, SDSSJO90745.0+24507, is going fast enough to leave the galaxy completely, which is quite unusual.

- A new neighboring star, at only 7.8 light-years away from us. This one, a small addition to the neighborhood at only 7 percent the mass of the Sun and 0.3 percent its luminosity, joins our other nearest neighbors: Barnard's star at 6 light-years away, and, nearest of all, Alpha Centauri, at 4 light-years. Welcome, 5025300.5+165258.

The main sequence stars, occupying most of the H-R diagram, present the basic way to understand it. At its clearest relationship, the diagram shows a star's surface temperature on the horizontal axis (with higher temperatures on the left, not the right); on the vertical axis lies the luminosity of the star, which proceeds brighter as one reads up, in units of 10 times the luminosity of our Sun. (Our star can be found toward the middle of the main sequence in Figure 12.3.)

The most luminous stars on the main sequence, found in the upper left, live the shortest lives. They are also not only the hottest but also the most massive. All main sequence stars feature similar nuclear "furnaces," steadily fusing hydrogen atoms into helium to release their energy. Their brightness depends on their mass since the greater the mass, the greater the inward force of gravity, which must be balanced by the force of the nuclear fusion engine in the core.

Energy and Color

Energy is emitted by stars in various forms known as the electromagnetic spectrum: gamma rays, x-rays, ultraviolet light, visible light, infrared light, and radio waves. Most main sequence stars radiate in all these forms of light, but they also tend to specialize in their emissions. The stars that emit a great deal of ultraviolet light are bluish, those emitting a lot of their energy as infrared radiation are reddish, and those blazing greatly in visible light are yellowish like the Sun. This is why color is featured on the H-R diagram. Star color is key to star temperature, too—bluish stars are the hottest, while those in the lower right corner of the H-R diagram, the reddish stars, are the coolest. Our bright yellow Sun is in the middle, as usual. (For a discussion of a star's spectrum, the spectral lines emitted by various elements within a star, see Chapter 18, Frontiers in Astronomy.)

Off the Main Sequence

Stars change their temperature and luminosity and thus move off the main sequence as they age. By that time they have consumed a great deal of their nuclear fuel, which alters them, though not immediately to cooler. These "elders"—the super giants and white dwarfs of the H-R diagram—and their procession toward death is described in the next chapter.

Major Star Types, Main Sequence

Before leaving the main sequence, the home of most active, living stars, it is worthwhile to profile some of the major types of stars occupying this zone:

- O stars (upper left on the H-R diagram) are massive (up to 100 times the mass of our Sun), hot (up to 50,000 degrees Kelvin), and short-lived ("only" about 1–2 million years). The fact that O stars even exist today shows that star formation has not ended entirely in the universe. These are dramatic flash-in-the-pan stars (which are, by the way, highly unlikely to have a family of planets since there is not enough time for planets to form before the star dies).

  Within the O star family, Wolf-Rayet stars live only about 100,000 years in this stage, a mere astronomical instant. As the cores of super massive stars which had previously lost their outer atmosphere in a stellar wind, Wolf-Rayets still give off very strong winds and intense radiation, lighting up the dust clouds of their birth into bright blue jewels. The hottest Wolf-Rayet found so far blazes at 120,000 degrees Kelvin, about 20 times hotter than our Sun. It will not be with us much longer.

- B stars (moving down the main sequence) live longer (about 10 million years) and are a bit cooler (about 20,000 degrees Kelvin at their surface).

- A stars last about 500 million years, blazing at about 8000 degrees Kelvin.

- F stars are even more long-lived (about 2.7 billion years) and have surface temperatures around 7000 degrees Kelvin.

- G stars, of which our Sun is one, live about 10 billion years (our Sun is projected to come in at about 10 billion or so, birth to death) and are 5000–6000 degrees Kelvin.

- K stars, dimmer and cooler than the Sun at about 4000 degrees Kelvin, live longer, to about 14 billion years of age.

- M stars are the coolest in temperature and can be as small as only 0.08 the mass of the Sun. They will last at least 200 billion years.

The farther down this list one proceeds, the less hot and bright the stars become, but the more stars there are in that category.

Open Clusters, Globular Clusters

Most star formation occurs in a group situation, as the dusty gas clouds lead to the birth of many stars at about the same time. The most common star nurseries are known as open clusters (looser) and globular clusters (more tightly packed with stars). Open clusters often do not last a long time in astronomical terms, and the more long-lived ones tend to be far out, at the galactic edge, where gravitational disruption is less strong. Globular clusters are as old as the galaxy, but they can also be disrupted and partially scattered over time. Clusters typically disperse as the mass of the rest of the galaxy gradually but steadily tugs on the stars at their edge. Since clusters have lost the gas from which they formed, they can no longer create new stars to take the places of those lost.

The open clusters, of which there are thousands in our galaxy, lie in the galactic plane, or disk. They are usually composed of a few thousand stars (or fewer) in an area about 10–30 light-years across. A star grouping easily visible with the naked eye from the Northern Hemisphere, the Pleiades, is an open cluster.

The globular clusters, of which there are only about 150 in the Milky Way, are more often in the galactic halo, the nimbus of gas and stars above and below the galactic plane. These clusters are balls of light composed of up to a million stars in a space about 60–150 light-years across. Gravity here moves the stars around like a 3-D pinball machine. Some of the stars fly out of the cluster entirely and join the rest of the galaxy. Over time, many do.

Far more unusual today, though probably not in the early universe, are the *superclusters*. Several of these have been detected, one of them in our galaxy at "only" 10,000 light-years away from here. In an area just 6 light-years across lie crowded stars which add up to the mass of about 100,000 Suns. (In our more immediate area the ordinary star density is, by contrast, about 10 solar masses in that volume of space.) The skies would be dazzling bright for anyone near this young supercluster. Another type of supercluster, a gathering of galaxies, not stars, is discussed in Chapter 14, Galaxy Formation and Development.

Globular clusters are collections of thousands of stars, held by their mutual gravitational attraction into a tight ball. This globular cluster, M4, contains several hundred thousand stars. NASA

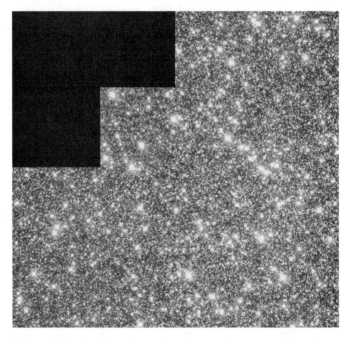

Omega Cen is a very large globular cluster. Here we see only a small portion of the cluster with seemingly countless stars in our field of view. In fact, the stellar density in massive clusters like Omega Cen is so high, that stars occasionally collide with one another. This almost never happens in most regions of the Milky Way. NASA

NGC 265 is a large open cluster in the Small Magellanic Cloud. Notice how open clusters do not show the compact, round shape of globular clusters. NASA

Variable Stars

Variable stars come in an immense variety of types and can be of almost any age and location on the H-R diagram. (Also, any star in the process of birth, building mass, is variable for a time; such stars are called the T Tauri Stars.) But, there are two important kinds of classic variable stars, both emitting pulsating light that varies from brighter to dimmer as they swell and shrink to some degree.

- *Cepheid variables.* These stars pulse every few days up to every few months (the brighter the star, the longer the cycle). The "North Star," Polaris, visible in the Northern Hemisphere from yards or parks away from city lights, is in this category. Variability in a star is quite significant because the relationship between its pulsating period and its luminosity, once known, allows astronomers to figure out its distance from us. With enough "mile-markers" like these Cepheids in place, we can judge the distances of other stars better and hence the extent of the universe itself. Astronomers call them *standard candles*.

- *RR Lyrae variables*. These stars have a variability that changes by the hour. With swelling and contracting this fast (and even as fast as the Cepheids do), it is no wonder that their location on the H-R diagram is known as the Instability Zone.

Other unusual stars, called the renegades, may be in terms of their mass/temperature/luminosity straight out of main sequence "central casting," not atypical in those ways. But they do seem to be escapees from the core of the galaxy, having galloped out into our Sun's general neighborhood (about halfway between the galactic core and the outer edge of the galaxy). These renegades are certainly not orbiting the center of the galaxy, remaining in their general birth neighborhood, the way most stars do. Instead, they are moving in more or less a straight line out away from the galactic center.

One runaway being studied is probably escaping from a too-close encounter with a powerful pair of binaries and is racing from them at 200,000 miles (321,869 kilometers) per hour. It is even lighting up an old gas cloud which it is passing through. This star is known as the *Flaming Star*. Observers should look now at AE Aurigae (its official name): in another 20,000 years, it won't be flaming anymore, having passed through the cloud.

Stars in Small Groups

Though it is simpler when discussing star evolution to focus on single stars with some attention to the close binaries and their atypical development, many, many stars in the universe are not going it alone, though their evolutions maybe separate. These stars are not necessarily in a relatively short-lived star cluster either. Actually quite common, they are the ordinary binaries and even triplet and quadruple stars, formed out of adjacent nodes in their original gas and dust clouds. Their percentage in the universe seems to be almost 50 percent.

What makes a star part of a pairing or a multiple system is mutual orbit. These motions can be quite complex. It would probably be unusual to see a full planetary system, one where life might evolve, around a binary, triplet, or quadruplet star. The gravity of the companion stars would have made any planetary orbits unstable.

Brown dwarfs, though they are not stars because their mass is not enough for nuclear ignition to have taken place, can indeed be part of a multiple system. This is especially possible if the small group of stars had emerged together from the same node. The objects which became the brown dwarfs were the ones which captured the smallest chunk of the raw material. (A brown dwarf is the size of 13–75 Jupiters.) Often there are even pairs of brown dwarfs in systems or groupings of stars. Less than the mass of 13 Jupiters, the object is considered a planet unless it has somehow ignited.

Whether found alone or in a multiple system, a brown dwarf usually has a surface temperature of 2000 degrees Kelvin or less, even as low as 800 degrees Kelvin (this is, amazingly, less than twice as hot as a hot oven in an Earth kitchen). They become even more dim as they age, and they tend to rotate slowly. This makes brown dwarfs a bit hard to find—the first one was found only about 10 years ago. But now we know that there are probably more of these "failed stars" in the universe as there are the real stars. Six brown dwarfs have been spotted in the process of planet-making, too; this is promising since their own modest luminosity makes the planetary disk easier to see. (Brown dwarfs are also discussed in Chapter 17, The Future of the Universe.)

Quick Tips

1. Binary stars are very common, especially those born out of the same gas and dust cloud at about the same time.
2. Open clusters are more common than globular clusters because gravity from outside the cluster gradually pulls it toward a looser density.
3. Variable stars pulsate from dimmer to brighter and back again at a rate that varies from every hour or so up to every few months.

Star Layers

A typical main sequence star like the Sun has several layers. The core, where the nuclear fusion occurs, is at the center. The next layer out is the radiative zone, through which the energy created in the core steadily radiates out. Surrounding that is the convection zone, where seething plumes of heat move the energy closer to the star's surface. From the thinner photosphere layer, above, the energy shines out as light (and all the forms of electromagnetic radiation). The chromosphere, as the outermost layer, lies just above the photosphere.

Surface and beyond surface features of a typical star such as the Sun include sunspots (magnetic storms), prominences (gas bursts that curl outward), the corona (the nimbus surrounding the star, which can be seen from Earth only in solar eclipses), and coronal holes and coronal streamers (through which the star sheds some mass). This shedding creates the stellar wind that sweeps out from the star. (For more on the stellar wind, and the Sun in general, see Chapter 8, The Sun's Role in Our Solar System.)

Not all main sequence stars have these features, and the proportion varies. Stars with higher mass than our Sun tend to have no convective zone below the photosphere. And lower mass stars tend to have thicker convection zones, ones that can reach all the way down to the core.

When a star leaves the main sequence, having exhausted the fuel in its core, the layering changes considerably. At that point, the star is on its death trajectory, the subject of the next chapter.

It's a Wrap

In this chapter you have learned that:

 The first generation of stars were massive, bright, and short-lived.

 Stars form as clumps within dust and gas clouds, enlarging as their gravitational attraction increases.

 Star matter is primarily hydrogen and helium which, through nuclear fusion, ignites the star.

 Stars are still forming today, but at not nearly the rate as that in the early universe.

 Mass, temperature, and luminosity are the parameters used to classify stars on the H-R diagram.

✔ Main sequence stars are far more common than the numbers of red giants, red supergiants, and white dwarfs on the H-R diagram.

✔ The larger the star, the more quickly it evolves.

✔ Stars are divided into O, B, A, F, G, K, and M categories, which describe their differing temperatures, brightness, and longevity.

✔ Tthe most common formation site for new stars is in an open cluster or a globular cluster.

✔ Variable stars pulsate in their light emissions.

✔ Rrenegade stars, also atypical, are not orbiting the galactic center.

✔ Main sequence star layers range from the core out to the chromosphere.

Test Yourself

True/False Questions

1. The dust and gas clouds within which stars are formed are quite hot.
2. Astronomers have seen the earliest two generations of stars.
3. The first generation of stars began to form the heavier elements found throughout the universe today.
4. In the universe today, the most common kind of star is low-mass.
5. Some stars are on a path to leave the galaxy.
6. The stars with the shortest lives of all are the Wolf-Rayet stars that live about 100,000 years.

Completion Questions

1. The first generation of stars in the universe were up to _____ times the mass of our Sun today.

2. At the star's birth it uses _____ and _____ to push away the remnants of the dust and gas cloud around it.

3. Most stars on the H-R diagram are _____ stars.

4. The main types of stars, classified by letter, and varying in heat, mass, and temperature are the _____ stars.

5. The two main types of variable stars are the _____ and _____ variables.

Multiple Choice Questions

1. The universe before the birth of stars is called
   a. The Big Bang
   b. The Dark Ages
   c. The H-R diagram

2. A protostar is
   a. A star that has not yet ignited
   b. The first star to be born within a gas and dust cloud
   c. The ideal or classic star on the H-R diagram

3. Stars are divided into
   a. High-mass, low-mass, and close binaries
   b. High-mass and low-mass
   c. Larger than the Sun and smaller than the Sun

4. Close binary stars can
   a. Form planet especially easily
   b. Be born very quickly
   c. Consume each other

5. A star burns out when
   a. It emits too much ultraviolet and infrared light
   b. Its nuclear fuel is exhausted
   c. It experiences a supernova explosion

6. Most stars are born in
   a. Close binaries
   b. Clusters
   c. The Sun's neighborhood of the galaxy

ANSWER KEY

True/False
1. F   2. T   3. T   4. T   5. T   6. T

Completion Questions
1. 150 times   2. Stellar winds and jets   3. Main sequence   4. O, B, A, F, G, K, and M   5. Cepheid and RR Lyrae

Multiple Choice Questions
1. b   2. a   3. a   4. c   5. a   6. b

CHAPTER 13

The Deaths of Stars

Do I Need to Read This Chapter? You should read this chapter if you need to review or learn about:

➡ How average main sequence stars become red giants or red supergiants

➡ The way these giants transform to white dwarfs

➡ The role of a supernova event

➡ How some high-mass stars move from their giant or supergiant phase into neutron stars

➡ The way some high-mass stars become black holes

➡ The role of supernova explosions in the death process of stars

➡ The death process for low-mass stars

Get Started

Even one step toward death for a star can be cataclysmic: for example, a neutron star named SGR 1806-20 recently flared brighter than all the stars in our Milky Way galaxy put together. This energetic emission, a short gamma ray burst of only about a single second, occurred less than 2 years ago.

Main Sequence Stars All Eventually Transform

Though the deaths of stars may seem sad, without events like these the existence of human beings would not be possible, neither would there be any form of life in the universe. Stars forge hydrogen into helium and then gradually create, by fusion, the heavier elements, too (much of chemistry itself, in fact). In one phase of their death throes—known as a supernova explosion, to be discussed later in the chapter—these elements are catapulted very suddenly out into the universe; they thereby become available for the next generation of stars to accrete and process further, elaborating more chemical elements with each generation of stars. These supernova events also cause immense shock waves, which ripple out through the gas and dust clouds of the area to jumpstart star formation in those places. So star death, in these and their other throes, is essential for life.

Main sequence stars constitute 90 percent of the universe's stars, since this is the main phase in which stars live their lives. (See Chapter 12, Star Formation, Star Development, for star birth and evolution as well as for an explanation and illustration of the main sequence.) Death begins for these stars because they eventually exhaust their nuclear (chemistry) **fuel** and can no longer create fusion. The more massive the star, the faster it moves toward death. In this section, we will first sketch the average death scenario, and then proceed to the special situations of very high-mass and low-mass stars.

As the typical main sequence star conducts its nuclear fusion, the helium gradually accumulates and the hydrogen in the core "furnace" gradually depletes. The heavier star elements, present in much smaller percentages, do not undergo fusion at this stage. (Helium itself fuses only at higher temperatures than that exist in the star's core at this point.) The more massive the star, the more quickly it burns out its fuel, even though it begins with more fuel.

Gradually gravity takes over as the energy supply diminishes and the core begins to contract. When a star shrinks under the action of gravity, it heats up. This new heat then allows the smaller zone of hydrogen outside of the core to begin fusion again. This energy is actually quite substantial, almost as though the star were burning anew to save its life. As this heat dissipates out into the outer layers, they begin to expand. The now-helium-rich core proceeds to contract further and the outer layers swell further. The star expands radically.

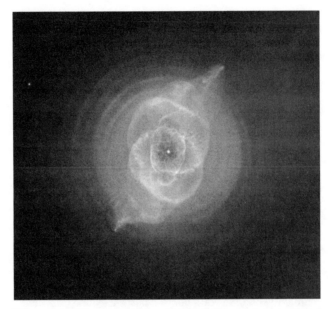

When a star like the Sun ages, it will swell up to become a red giant and blow off some of its mass into space. When the outer layers of the Sun are gone, the extremely hot core down inside becomes exposed and lights up the gas and dust thrown off as a red giant. This produces a planetary nebula like the Cat's Eye Nebula shown here. NASA

The star is now in its first quite obvious stage of death. It becomes a red giant if its original mass was low or medium in size, and a red supergiant if its life was as that of a high-mass star. (The lowest mass stars have not yet undergone this transformation anywhere in the universe—the universe is simply not old enough yet!) Red dwarfs, though, as seen on the H-R diagram, are not dying stars. Instead, they are simply small, dim, reddish stars right on the main sequence.

Red giants and red supergiants are both huge and bloated. The red giants are brighter than their former selves because of the new vastness of their light-emitting surface, but they are also cooler since they are both diffused and fuel-inefficient in a nuclear sense. The difference in size between these two kinds of stars is significant: if a supergiant were centered on the Sun's position in our solar system now, it would expand out past, say, the orbit of Mars or Jupiter. A red giant, however, would "swallow" only the now-existing orbits of, perhaps, Venus, Earth, or Mars. Both these distended states are only the first stage of star death. (How this affects planets belonging to these stars, if any, is discussed later in the chapter.)

Three of the largest supergiants detected so far are KW Sagittarii, V354 Cephei, and KY Cygni. If centered on our Sun, the outer edges of all three would extend nearly as far out as the orbit of Saturn. They are too far away to be seen by the naked eye from Earth, though they are immensely bright.

The red giants Betelgeuse and Antares, though they are somewhat smaller than these prizewinners, can, however, be seen right from a yard or park in the Northern Hemisphere, often even in the presence of some city lighting. Not even a pair of binoculars is needed. These stars could collapse, entering their next stage of star death anytime between tomorrow and a few thousand years from now. Take a look at them, now.

Quick Tips

1. Main sequence stars begin death by exhausting their core hydrogen fuel.

2. Core shrinkage causes heat, which allows noncore hydrogen to re-engage the nuclear furnace, a stage that is relatively short-lived, but dramatic.

3. The death process at this point causes the star's outer layers to expand exponentially, transforming the star into a red giant or red supergiant.

Our own star will not escape death. In fact, the Sun will become a red giant in 5 billion years.

Problems for any solar system inhabitants at that time will begin well before this full bloat stage, however. As soon as 2 billion years from now, the Sun will be in its earliest stages of swelling. Heat reaching the Earth will become enough to evaporate the oceans. Land surface temperatures will become increasingly scorching.

As the Sun approaches the full red giant stage, in about 3–4 billion years, our planet's temperature will surge hotter than 750 degrees Celsius. This will eliminate most life here, though some life forms such as insects and "extremophiles" (primitive bacteria-like creatures which inhabit extreme environments) could probably survive.

As the Sun continues to shrink inside its bloated shell, it will become hot enough to spark the fusion of helium. After shrinking for awhile, it will begin another stage of swelling. At this mature red giant stage, the Sun will engulf the inner planets out through the Earth's orbit. Inside the diffused shell, helium will be fusing into carbon, but no other fusion will be able to take place since the active engine will no longer be massive enough. The ultimate end of our star will be as a carbon white dwarf star.

Mercury will be vaporized by this time, but Venus and Earth will survive as objects for awhile longer. This is because they will no longer be in their present orbits. Since the Sun will have been shedding and expelling materials gradually but steadily, it will be less massive. With less gravitational force to hold Venus and Earth in orbit, these two planets will withdraw slightly. Earth will become not quite twice as far from the center point of the Sun as it is today.

Unfortunately, that will not be good enough for even the hardiest life forms on our planet. The Sun will also be about 5000 times brighter than it is now. This will heat the surface of the Earth to 1300 degrees Celsius, and our rocks will be transformed back to their molten form.

The Giants Next Become White Dwarfs

The "blown-up-balloon" stage of a star's death, though long in human terms, is a short and unstable period in terms of the long life of a star—less than a few hundred million years. After this period, an average-mass star begins to shed its expanded outer layers, leaving only a hot, collapsed core. When nuclear fusion is over, these stars cool and dim over time. Once shrunken and faint, they become white dwarfs.

A typical white dwarf may be no larger than the Earth. It is, however, incredibly dense with star material, perhaps with the mass of our Sun packed into an Earth-sized package. A teaspoon of it would weigh (about 50) **several** tons. The first white dwarf ever discovered is the close companion of the very bright star Sirius, easy to see with the naked eye (which, though, makes the two stars appear to be one); look for it near the horizon in Northern Hemisphere summertime from a yard or park.

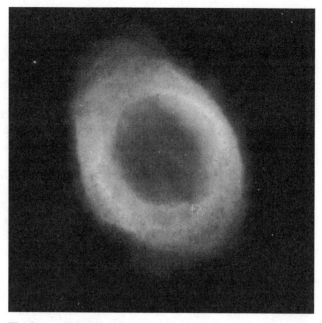

The famous Ring Nebula is a good example of a planetary nebula. It is round in shape and this confused early astronomers into thinking they had found a new planet in our Solar System, hence the somewhat confusing name. Notice the faint star at the center, it will become a white dwarf star. NASA

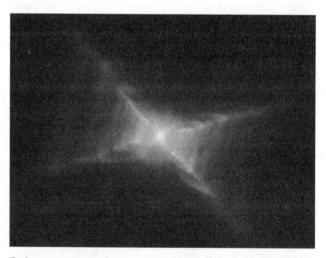

Before a star produces a full fledged planetary nebula, it makes the transition from a red giant star to a white dwarf in a relatively short time. This short lived phase can produce some very strange looking nebulae, such as the Red Rectangle shown here. Outflows of gas and dust are ejected from the star in two opposing directions, producing a shape like two ice-cream cones touching at their tips. NASA

The condensed mass of a white dwarf can act quite powerfully, especially in certain configurations. Astronomers have found a binary pair of white dwarfs orbiting each other so closely that both could fit into the space between orbits of the Earth and the Moon. This is tighter than any two astronomical objects ever discovered (although binary pairs of neutron stars could theoretically be even closer). The orbital energy, shed as the two stars pull each other even closer and faster through their gravity, is creating gravitational waves; these waves are among the strongest such sources the our galaxy. This Einsteinian spacetime phenomenon is described in Chapter 9, Origins of the Puzzles of Spacetime and Gravity.

After a long period as white dwarfs, many of the stars (those which are not close binaries) eventually will cool into black dwarfs, fading out of sight completely. Occasionally, however, a white dwarf will explode in a supernova event known as a Type 1a. Another of the several supernova variants is mentioned in the next section.

Quick Tip

The next stage of life/death for an average main sequence star is to become a white dwarf.

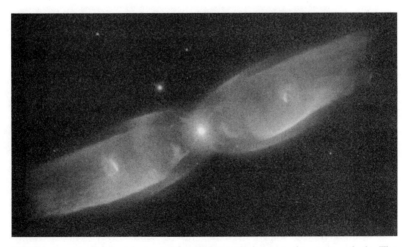

M2-9 is another object in transition from red giant to planetary nebula. The bipolar nature of the outflow from the aging central star visible at the waist of the nebula is strikingly apparent in the bow-tie pattern it produces. NASA

Peering deep into the globular cluster M4, the Hubble Space Telescope can see a myriad of faint stars. These are White Dwarf stars, the dead remnants of stars like the Sun who have lived out their lives and are no longer undergoing nuclear fusion in their core. NASA

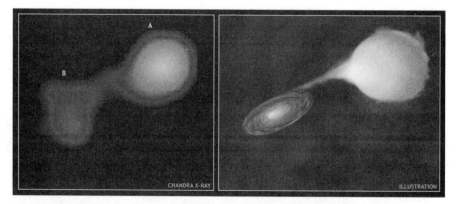

When two stars in a binary age together, the more massive star will evolve faster and become a White Dwarf first. Later the lower mass star will swell up into a Red Giant and transfer some of its mass onto the White Dwarf. In this Chandra X-Ray Observatory image, we see the red giant Mira and its compact White Dwarf companion. An artist's conceptual drawing illustrates how gas flows from the huge red giant down onto the tiny white dwarf. NASA

The Role of Supernova Events

For a star to move from the red giant phase to become a white dwarf can be as gradual as the process of steadily shedding its outer layers (as mentioned in the previous section). But for red supergiants, which emerge from massive stars, the end can be more dramatic: an explosion in which chunks and whole layers of the star are stripped off, blasting out star material and sometimes gamma rays. This event is called a supernova. Supernovas are known to occur in several scenarios. The transition of a massive star with hydrogen remaining as it transforms into a neutron star, called a Type II or core-collapse supernova, is one of these. (Some of the others are discussed later in the chapter.)

In a supernova, the explosion throws the outer layers of the star off into space. This action forms a shell of gas that expands away from the star, very rapidly for the first few decades, then more slowly after that. Shock waves from the original explosion travel out through the gas, shaping this diffused material in complex ways. For a brief few weeks, a supernova event at this stage can outshine a billion Suns. Astronomers can see the gaseous shells created, since they are lit by the energy of the blast and sometimes by a neutron star left inside. A shell can remain in place for thousands of years until it gradually dissipates into interstellar space. These supernova shells can be seen in visible light, polarized light, or by satellites that observe x-rays, infrared light, or another dimension of electromagnetic radiation. One such supernova, known as 1987 A, has been observed consistently since 1999. The shell is still so thick that the object inside is shrouded, though astronomers believe it to be a neutron star whose core simply collapsed after the supernova explosion.

These often exquisite star shells are called nebulas or supernova remnants. Dramatic objects, they outshine the star which created them in the high-energy x-ray part of the spectrum.

Supernovas occur only one to three times every 100 years within the Milky Way. But they are, by definition, so bright that they can be observed occurring even in galaxies very far away. Two supernovas visible now lie at 80 million light-years and 440 million light-years from us.

Astronomers study supernovas to learn about a host of astronomical phenomena: the dust clouds existing in the area of space before the blast; the cosmic ray ions that proceed out from a supernova and hit our planet; the possible source of some burst of gamma rays; the expansion of the universe; and, of course, the nature of the exploded star itself.

Besides the Type 2 supernova, Type 1a has also been studied. This type of supernova occurs when a white dwarf in a binary star system receives too much mass from its companion, reaches a critical mass, and undergoes a rapid collapse. The collapse causes the explosion.

A grab bag of supernovas have been observed throughout the history of astronomy. Chinese observers recorded one in the year 1054. One was seen by Tycho Brahe in 1572 and another in 1604 by Johannes Kepler. Kepler observed the star with the naked eye as it appeared suddenly and unexpectedly in the sky, then brightened and faded over a year. He used this phenomena to conclude that the heavens were not unchangeable, a major shift in the thinking of his period. The gas shell of this star is still visible today.

Observers in the Southern Hemisphere must have noticed another supernova in the year 1006, one which is also still being studied today. Its shell not only remains, but changes in it are being observed even now. Nebulas can be molded by gravity or shock waves from any source they encounter as they gradually expand outward farther and farther from the point of explosion. Extrapolation from the brightness and distance of this early supernova has let astronomers know that it was originally brighter than Venus and about half as bright in the night sky as the full Moon. No written records exist of this 1006 event.

A "fake supernova" is also part of the array. A star that had been known as supernova SN1954J turned out not to have been blasted to smithereens in an explosion at all. Apparently, it had merely shuddered violently, as some extremely massive stars can do, yet it remains whole. For now.

Quick Tips

1. Supernovas are of several types since different death scenarios proceed in different grouping of stars or within single stars of differing masses and compositions.

2. The shells created by their blasted-out material, called nebulas, can be observed and studied up to centuries later.

3. Not every star experiences a supernova explosion by any means. Our modest-sized Sun almost certainly will not experience one.

A **less energetic nova-like** (supernova) event on star V838 Monocerotis has led to an unusual observation, of what is called a "light echo." This is light bouncing off the dust in an area's pre-existing dust and gas cloud. The (supernova) outburst, **observed** in 2002, was especially dramatic. The star puffed out to more than 800 times the size of our Sun and, for a few days was the brightest object in our entire galaxy.

This dying star, V838 Mon, underwent a Nova-like outburst in 2002. As it brightened, it illuminated the surrounding dust left over from previous outbursts, producing an ever changing image astronomers call a 'light echo'. NASA

High-Mass Giants and Supergiants Can Become Neutron Stars

Star death throes continue into further dramatic phases, especially for high-mass stars—those about 10–25 times the mass of the Sun or more. One of these is transformation into a neutron star.

One of the most massive and luminous stars known, Eta Carina is nearing the end of its dramatic, but short life. Eta Carina was the site of a giant outburst about 150 years ago, when it became one of the brightest stars in the southern sky. Massive stars like Eta live very short lives and explode as a supernova. NASA

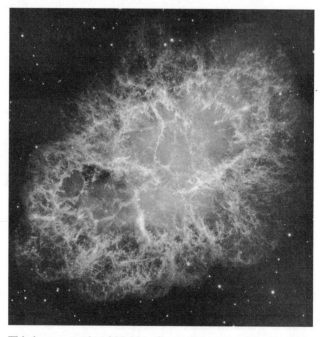

This large mosaic of Hubble Space Telescope images shows the Crab Nebula in incredible detail. The Crab is the result of the most energetic event in the life of a massive star, a supernova. Japanese and Chinese astronomers recorded this violent event nearly 1,000 years ago in 1054. NASA

In a supergiant, helium fusion continues to a certain degree even in the bloated stage, though the star also sheds mass at a furious rate. The core gradually compresses until the carbon within it (in smaller amounts all along) also ignites. Other elements, even up to the weight of iron, begin increasingly to fuse as the core becomes hotter and hotter. This is, in fact, how the heaviest elements in the universe have been created. As these heavier elements are expelled in star deaths, they become available for the next generation of stars.

> The supergiant Eta Carinae, which is at this stage of star death process right now, is expelling its mass at up to 621 miles (1000 kilometers) per second. Still one of the most massive stars in the galaxy, it emits more light in 6 seconds than the Sun does in a year. Nitrogen and other heavier elements are being blown out, and they are "seeding" the star's neighborhood, providing the raw material for the next stars. This amazing star, visible only from the Earth's Southern Hemisphere, is now 4 million times brighter than our Sun. It will explode as a supernova relatively soon, in astronomical terms.

As the heat in the star's core increases, the core can actually implode in other types of supernova explosions (a Type 1b, 1c, or Type II supernova, depending on the specific circumstances). Core destruction emits an immense amount of the most powerful form of radiation detected in the universe: a gamma ray burst. It is usually accompanied by x-ray emission. The total energy of the burst can outshine 1 billion Suns. In fact, this flush of radiation is typically how these types of supernova are detected. The long bursts can last several seconds up to several minutes, before fading completely over a few minutes. They have been associated even with the death throes of high-mass star that occurred 12.8 billion years ago—in other words, in the era of the first generation of stars. One such "gamma ray burst (LEAVE THE ER)," called GRB 050904, was detected out 12.8 billion light-years away from us, the farthest ever, and providing a glimpse back to when the universe was less than a billion years old. And another—close at 440 million light-years away—seems to have shown astronomers a burst of extraordinary length, about a half an hour. (Short gamma ray bursts originate in a different scenario and are discussed later in the chapter.) Both kinds of gamma ray bursts, called GRBs, are an important area of astronomical research, since they are so powerful and may turn out to be created in ways besides the explosions of massive stars.

The whole process of implosion and burst typically takes only a few seconds, though for about a week longer the star may be as bright as 10 billion Suns. Often, a highly dense object remains after the implosion, a neutron star. In it, the protons and electrons have fused to form neutrons (another type of elementary particle). This makes neutron stars the densest objects in the universe, each one like a giant single atomic nucleus held together by immense gravity. A star of about 12 miles in diameter can weigh as much as a couple of Suns. Anyone landing on a neutron star would also be flattened instantly into subatomic particles, and the escape velocity (to get away) would be almost half the speed of light. Sometimes gravity is so strong that a neutron star cannot form at all. A black hole is created instead.

> A new and very unusual type of neutron star was first discovered in 2006. Called RRATS (rotating radio transients), these stars emit powerful bursts of radiowave radiation—and in bursts that last a fraction of a second and occur every few minutes up to every few hours. In between, they are completely dark in every way.

Pulsars

A core implosion can also result in an especially fast rotation. This type of speeded-up neutron star is called a pulsar. Rotations can be close to 800 star spins per second, even though the mass is typically that of only about $1^1/_2$ Suns. The fastest of all, called millisecond pulsars, usually are part of a binary system with a red giant star transferring mass to the pulsar. The energy of the spin accelerates charged particles, which radiate light. Pulsars all eventually do slow down, though, as they radiate away their energy. Some 1500 pulsars have been found and studied so far, a few of them even in binary pairs. More pulsars are being found all the time, too.

> Given the immense energy of a gamma ray burst, long or short, or even sometimes weak, it is fortunate that the Earth's atmosphere stops these rays. Gamma ray bursts are so energetic that they can reach us from billions of light-years away, which means that we may typically be bathed in them from all directions. Actually encountering radiation like this unprotected would be fatal to a human being.

In addition to their spins, pulsars also exhibit extreme speeds, tearing through space at up to 680 miles (1094 kilometers) per second. Sometimes the paths are anomalous, too, and these pulsars are called runaways. The runaways may be neutron stars leftover from a supernova explosion that took place off center inside the massive, original star.

Pulsars typically emit radio waves too, along with other electromagnetic radiation, in two powerful "jets" or beams which sweep around like searchlights from a used car lot. In a pulsar named Vela, the x-rays in its jets are being accelerated by a voltage with the strength of 100 million lightening bolts on Earth. Why these beams differ among pulsars is not understood.

The luminosity of the Crab Nebula is powered by one of the most bizarre objects in nature, a rapidly rotating neutron star, or pulsar. Here we see X-Ray and Optical images combined to show the pulsar (the star in the center) and the highly sculpted gas surrounding it. NASA

Magnetar

Another kind of neutron star, formed in the death trajectory of a high-mass star, is called a magnetar. So far, less than 20 of them have been discovered in the universe. The magnetic fields of these stars are so strong that standing 621 miles (1000 kilometers) from one would kill a human by warping the very atoms in the body.

One magnetar, called SGR 1806-20, blasted us with short gamma rays and x-rays in December, 2004, from about 50,000 light-years away. This star, a tangle of magnetic field lines under a crust, is probably only about a dozen miles across and rotates every 7 or 8 seconds. This twists the magnetic lines into extreme instability. Finally, its crust snapped in what is sometimes called a "starquake." That caused the blast detected on Earth. It is the speed of rotation and the strength of the magnetic field that distinguishes the faster magnetars from more typical neutron stars. And SGR 1806-20's rotation has slowed only slightly since this energetic emission, though magnetars do tend to be short-lived.

Neutron Star Mergers

A merger of dying stars of this type, is, obviously, cataclysmic. What we detect on Earth from such an event is a short gamma ray burst. One such blast was measured at 0.03 seconds (the longest burst still considered a short gamma ray burst is a full 2 seconds long). The origin of a blast like this seems to be either the collision and merger of a binary pair of neutron stars, or, less frequently, the merger of a single neutron star and a black hole. (The second of these scenarios, more like a swallowing than a merger, is described later in the chapter when black holes are discussed.)

The too-close approach of two neutron stars in a binary pair is thought to take at least a million years, as they tug on each other until their orbits decay enough for them to merge. This kind of dangerous dance often occurs deep within a globular cluster, where star populations are dense. Since most binary stars are not neutron stars at all, it is amazing that several of these double-neutron star events were detected in one year, 2005. Their energy, at least, makes them noticeable.

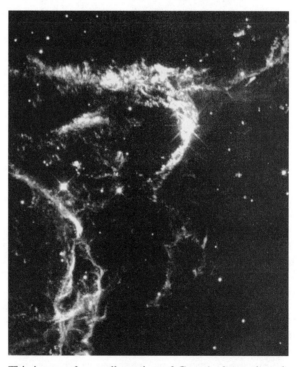

This image of a small portion of Cass A shows thread-like filaments that are regions of hot, glowing gas produced by the blast of a supernova explosion. As the blast wave travels out into space, it collides with the surrounding interstellar matter causing it to glow. NASA

Quick Tips

1. High-mass stars often move from the giant stage to the neutron star stage. In it, heavier elements undergo fusion.

2. Some neutron stars undergo core implosion and collapse, in one type of supernova, emitting energetic, long gamma ray bursts and becoming the densest objects in the universe.

3. Neutron stars that rotate especially quickly are called pulsars and millisecond pulsars.

4. Neutron stars that are unstable magnetically are called magnetars.

5. The merger of two neutron stars creates short gamma ray bursts and sometimes a black hole.

When a pair of neutron stars collides in a flame of gamma ray bursts, the result can even be the creation of a black hole, as mentioned above. Einstein theorized that gravitational waves would be created, too, bending spacetime. This is also held to occur if one neutron star collides with an already existing black hole and falls into it. An event like this seems to create x-ray emission as powerful as the gamma rays: an emphatic "birth cry."

High-Mass Giants and Supergiants Can Ultimately Become Black Holes

A black hole is truly the end of the line for a star. This avenue to nothingness, though a very powerful nothingness, is the fate of stars which lived life on the main sequence at the highest mass level, progressed to becoming red supergiants, collapsed not into neutron stars but rather experienced gravity as a runaway force. Gravity waxed so strong in these objects that, as the star compressed, not even light could escape it.

The black hole experience, as it affects individual stars, seem to include those of about 25–40 original solar masses and with cores of at least three solar masses at the time the process begins. (Stars belong to the 10–25 solar mass level are more likely to become, and remain, neutron stars at the end of their lives, as seen earlier in the chapter, though they can also be quite dramatic, producing supernovas.) Gamma ray bursts are emitted, immense jets of star material are expelled, and they glow in electromagnetic radiation from x-rays to radio waves. These death cries begin about 1 hour after the star's core collapses and continue for up to several months. Anything within a light-year of an event like this could become involved in it.

> Astronomers are now on the verge of actually seeing the shadow produced by the supermassive black hole at the center of our galaxy. The evidence in the light emitted by hot gas as it swirls into the black hole just before being lost forever. Only radio telescopes can penetrate the thick, obscuring dust between the Earth and the galactic core. As the light from the far side of the black hole is bent by the strong gravity near the event horizon, a "shadow" is created. Our black hole itself, as measured by the size of its event horizon, is expected to be smaller than the orbit of Mercury, though its mass and power are immense.

New black holes come in various sizes, and these sizes can change. Originally, their mass is determined largely by the mass of the original star. Some then condense to be no bigger in extent than the average asteroid in our solar system (though much, much more massive). Other black holes are able to consume more matter, their "food" coming either from accretion of nearby material (which can form a spinning disk like the one at our galactic center) or, far less often, from the merger of two galaxies' central black holes. The latter ones become far, far larger.

A typical galactic center black hole is very massive, since the centers of galaxies are well populated by stars and dusty gas clouds for it to gradually consume. They range from about 1 million solar masses to the largest ever detected, at the 3–4 billion solar mass level. (These "supermassive black holes" at galactic centers, detectable at even 13 billion light-years away, arealso discussed in the next chapter, Chapter 14, Galaxy Formation and Development.) Perhaps even stranger, a galactic center like our own Milky Way's may have several average-sized black holes (about 5–20 solar masses apiece) in orbit around the supermassive one. They are almost certain to fall into it eventually, hors d'oeuvres into a greedy party guest's mouth. In another galaxy, Markarian 766, hot blobs about the size of the Sun are in orbit around the central black hole, careening around it at the extreme speed of 20,505 miles (33,000 kilometers) per second. And in the Andromeda Galaxy—which can be seen from Earth with the naked eye far from city lights—some 200 new stars are swirling close and fast around the galactic center black hole. All these hapless astronomical objects will be joining their black hole, and probably soon.

At the other end of the size spectrum there may be extremely tiny black holes called pinpoint black holes. If they indeed exist, they would be smaller than an atomic particle but as massive as a city bus. Pinpoint black holes would not last very long, however, for empty space is not truly empty. It is a swirling froth of particle and antiparticle pairs that wink in and out of existence. Gravity near the event horizon of a pinpoint black hole would be so strong that it could drag one of these particles in, as the other escapes. The smaller a black hole, the more quickly it would evaporate in an interaction like this.

The massive black holes leftover from supernovas as well as the supermassive black holes at the centers of galaxies (to be discussed in chapter 14, Galaxy Formation and Development), though, will last much longer—many times the age of the universe— before they evaporate. They will continue to consume nearby gas and stars and grow with time rather than shrink, the way their pinpoint "cousins" do.

Especially alarming to anything in its path is yet another kind of black hole, the runaway. One of these has been seen dragging a companion star along at the breakneck speed of 248,548 miles (400,000 kilometers) per hour. It is moving outward from the galactic center, which means that the pair will eventually leave the Milky Way. This duo, named GRO J1655-40, was probably ejected from the galactic center by an especially powerful supernova explosion, which was also a bit asymmetrical. This blew the new black hole off course. Though this object is indeed in our Milky Way galaxy, it will, fortunately, not travel anywhere near to the Earth.

Quick Tips

1. Black holes seem to form in three main ways: a supernova explosion ending the life of a star, which was originally 25–49 times the mass of the Sun; a collision of binary neutron stars; a merger between an existing black hole and a neutron star.

2. The black holes near the centers of galaxies continue to grow larger and larger as they swallow material from those crowded zones. The largest one detected has the mass of about 3.7 billion Suns.

3. The gravity of a black hole is so strong that light cannot escape, though radiation is emitted near their event horizons.

A black hole's nature is clearly not typical of any other astronomical object's. Its edge, the event horizon mentioned above, is more like a point of no return than an edge made of matter. Here, gravity is so strong that the escape velocity would be faster than the speed of light. Since nothing can move faster than the speed of light, nothing escapes. Gravity this powerful actually warps spacetime. Inside the black hole, at the center, is a particularly strange feature called a singularity. This tiny point will not be well understood until astronomers understand how the theories of gravity and of quantum mechanics fit together. (For more on these basic theories of nature, see Chapter 10, Down to the Atomic and Subatomic Levels.)

> Most people are curious about "death-by-black-hole." Step within an astronomical unit of a massive one to attempt to find out. The powerful gravity will first pull the feet, stretching a human body to a strand of spaghetti. At the same time, its force will begin to compress your shoulders together. In an instant to you, but an eternity to someone looking from far away, the end will come.

Black holes, also strangely, can seem to emit jets of x-ray radiation in the process of consuming material. Actually, it is the hot gasses spiraling down toward the black hole that emit the light, not the black hole itself. A black hole named Cygnus X-1, about the mass of 10 Suns, is slowly pulling at a nearby supergiant and consistently spitting out the brightness of 100,000 Suns as it does so.

Death for Low-Mass Stars

Slow and steady may win the race. Low-mass stars live long lives on the main sequence, fusing hydrogen into helium; this energy moves outward from their cores in radiative diffusion, then farther out through the cooler upper layers via convection (the kind of plume-like effect one can see in boiling soup). The lower the mass of the star, the thicker is this steady convection zone. The lowest mass stars can live this way for tens of billions of years.

Gradually, however, even these stars begin to exhaust their fuel. Without that outward press of energy, gravity takes over. The core of the star begins to compress and the outer layers puff out. The result is a subgiant star, much smaller than the supergiant stage of a high-mass star.

After another billion years or so, the star expands more and succeeds in becoming a red giant. The core, made up entirely of helium, gradually heats, through gravity, to about 100 million degrees Kelvin. At the same time, the star's stellar wind gradually expels the outer shell, without any supernova explosion. And the star becomes a helium white dwarf, the end for a very low-mass star.

Very Last Stages

Stars like the Sun contribute something else to space also. For a million years, their slowly leaked star material glows, creating one of the most beautiful astronomical objects of all—a planetary nebula. (An old name, it may be confusing since these nebulas have nothing to do with planets.) Our galaxy contains tens of thousands of these, with a light-year in diameter as the typical size.

Finally, a nebula like this dissipates. The solar mass star is now simply a white dwarf. Unless this cool, small object becomes part of a binary, or develops unusually in another way, its life has essentially ended. When it has cooled enough that it emits no light at all, it becomes a black dwarf.

The formation of new stars in the universe will probably not go on forever. Even though dying stars emit a great deal of material, which provides raw star stuff for the next generation, more and more of this material becomes "locked up" in star corpses like those discussed in this chapter. In our Milky Way Galaxy, about seven new stars seem to be born every year and only about two die in every century. A ratio like this seems to be typical over the universe.

 In this chapter you have learned that:

✔ The death trajectory of a star depends crucially upon its initial mass, with the death stages of high-mass stars the most dramatic.

✔ Our Sun's death, that of an average star, will include a vastly expanded red giant stage, which will ultimately end life on our planet, in 5–6 billion years.

✔ High-mass stars can undergo various types of supernova explosions, which move them to the next stage of death.

✔ The exhausting of a star's nuclear fuel is the core reason for its death.

High-mass stars become red giants and red supergiants, then either neutron stars or black holes.

 Low-mass stars become white dwarfs, then black dwarfs.

 Though stars all die, new stars are being born all the time—and out of the heavier elements forged in the earlier generation stars, as well as from the remainder of the primordial hydrogen and helium of interstellar space.

Test Yourself

True/False Questions

1. Stars fuse hydrogen and helium into energy throughout most of their lives.
2. Red giants and supergiants are much brighter than they were before the first stage off the main sequence.
3. Pulsars and magnetars are types of neutron stars.
4. Binary stars can exchange mass and shed gravitational waves.
5. Supernova explosions happen only to high-mass stars.
6. Pulsars are the fastest stars in rotation speed.
7. A dying star can be as small as 12 miles in diameter.
6. Black holes are most often the death stage of stars with about 10 original solar masses.

Completion Questions

1. After their lives on the main sequence of the H-R diagram, stars become either _____ or _____ or _____ next, depending upon their mass.
2. Stars that lived as average-mass stars will end their lives as _____.
3. Stars that lived as high-mass stars will end their lives as_____ or _____.
4. Magnetars have the strongest _____ of any star.
5. The merger of two neutron stars or of a neutron star and a black hole can create a_____.
6. Objects that probably will not escape a black hole are those within _____ light-year of it.
7. The largest black hole detected has the mass of _____ Suns.
8. The center of a black hole is called a_____.

Multiple Choice Questions

1. High-mass stars
   a. Die more quickly than low-mass stars
   b. Have more hydrogen-helium fuel to begin with
   c. Both

2. When the Sun becomes a red giant, it will
   a. Swallow the Earth
   b. Boil the oceans
   c. Draw the Earth's orbit closer to itself

3. Neutron stars
   a. Are the densest stars
   b. Emit shot gamma ray bursts
   c. Generally develop into black holes

4. An event horizon is
   a. The beginning of a low-mass star's death
   b. The discovery of gravitational waves by astronomers
   c. The energetic edge of a black hole

5. Black holes can
   a. Be the size of an asteroid
   b. Be the size of a galaxy
   c. Move between galaxies

6. In black holes
   a. Gravity prevents the escape of just about everything
   b. Sound waves are emitted
   c. Both

7. Low-mass stars
   a. Live the longest
   b. Experience only one supernova
   c. Have planets in their nebulas

ANSWER KEY

True/False
1. T 2. T 3. T 4. T 5. F 6. T 7. T 8 F

Completion Questions
1. Giants, supergiants, or subgiants 2. White dwarfs 3. Neutron stars or black holes 4. Magnetic fields 5. Short gamma ray burst 6. 1 7. 3.7 billion, 8. Singularity

Multiple Choice Questions
1. c 2. b 3. a 4. c 5. a 6. c 7. a

CHAPTER 14

◆◆◆◆◆◆◆◆◆◆◆◆◆◆◆◆◆◆◆◆

Galaxy Formation and Development

◆◆◆◆◆◆◆◆◆◆◆◆◆◆◆◆◆◆◆◆

Do I Need to Read This Chapter?

You should read this chapter if you need to review or learn about:

➡ The three main types of galaxies

➡ The Milky Way's formation and Earth's position within its spiral arm system

➡ The Milky Way's disks, central bulge of stars, and black hole

➡ The Milky Way's interstellar medium, star clusters, and halo

➡ Ecology and evolution of the Milky Way

➡ Andromeda Galaxy

➡ Our Local Group and Local Cluster of Galaxies

➡ Our Local Supercluster and the Great Attractor

➡ Quasars

➡ Interactions of galaxies in space

➡ Relationships among galaxies across time

Three Main Types of Galaxies

The three main types of galaxies are:

- Spiral galaxies. These galaxies have a central bulge of stars, spiral arms, and at least one disk of darker, cool gas and dust, which slices though the middle on the horizontal axis, more of this material also dispersed throughout the galaxy, star clusters, and a halo surrounding the system. Stars are concentrated heavily in the central bulge, quite concentrated in the disk, and lightly dispersed in the halo.

  Within this grouping of galaxies, about a third are known as barred spirals; these have a thick bar of stars across their central axis, with the spiral arms proceeding from the ends of the bar. (The Milky Way seems to have a bar, though it is not particularly distinct.) Another subcategory of spirals is the lenticular spiral, which mostly lacks spiral arms. Spiral galaxies together comprise more than 75 percent of the (full-sized) **large** galaxies of the universe, and they tend to be more massive than other types of galaxies, except for the giant ellipticals mentioned below. Their detectable diameters range from about 20,000 to 100,000 light-years.

- Elliptical galaxies. Lacking outer arms and usually even inner disks, these galaxies look like immense eggs (if they are large) or loose clusters (if they are small). The giant elliptical category comprises most of the largest galaxies in the universe.

  Ellipticals vary considerably in both mass and diameter. In them, the stars' paths move every which way, unlike the more orderly rotations around the galactic center seen in spirals. Though stars are more concentrated toward the center than in the halo, the differences in concentration are not nearly as pronounced as in spiral galaxies. The gas and dust clouds are also sparse in these galaxies but are hotter in temperature than in the spirals. Ellipticals are usually smaller than the average spiral galaxy and are generally found within

Seen face-on, M101 is a typical spiral galaxy.

NGC 1300 is also a spiral galaxy, but has a prominent bar structure through the nucleus. NASA

clusters of galaxies. (Only about 15 percent of them are outside of clusters.) This tells astronomers something about their evolution, as is mentioned later in the chapter. Probably the most common type of galaxy in the universe, across all sizes, is the dwarf elliptical; some of these have only several million stars, and a few are home to less than 100,000 stars.

M32 is a small, nearby elliptical galaxy. NASA

- Irregular galaxies. These galaxies were more common in the early universe than they are now. Today, they comprise the smallest category of galaxy. Irregulars tend to look disorganized even when they are massive and undergoing considerable star formation; the reason for this is discussed later in the chapter.

This overview considers only the basic, detectable elements of galaxies, the stars, and the gas and dust clouds. As described in the next two chapters, dark matter and dark energy also seem to be present in every galaxy, and in great amounts.

Composite of multi-wavelength images of the active galaxy M82 from the three Great Observatories: Hubble Space Telescope, Chandra X-Ray Observatory, and Spitzer Space Telescope. M82 is an irregular galaxy that is being disturbed by intense star formation taking place in its nucleus. NASA

Galaxies can also be differentiated for research purposes by their luminosities, namely, the predominant type of light they emit. Ultraviolet galaxies, for example, are rich in young, massive, hot stars (more than 10,000 degrees Kelvin hot), found in a diffuse outer disk. Much of the star formation in these areas seems to be set off when two galaxies entangle their outer arms or interact even more substantially. Some of these star nurseries, however, may, instead be in the primordial gas clouds; these clouds never formed stars earlier and dates back to the very early days of the universe. In general, galaxies contain their oldest stars in the central bulge and their youngest on the edges.

An energetic galaxy in the active star formation phase may radiate more than 20 billion times as much energy in the ultraviolet alone than our Sun does across all its wavelengths.

Galaxies also vary considerably in size. An ordinary dwarf galaxy may contain only 100 million stars, some even far less. A giant galaxy is home to at least 1 trillion of them. The Milky Way, considered large with its 100 billion stars in an area more than 100,000 light-years across, is somewhere in between, far larger than nearby dwarfs, though not as large as a giant elliptical.

Elliptical galaxies, once though to be past the stage of star formation, are now known to produce new stars. They do, however, contain mostly older stars. Galaxies, in general, seem to build from the inside (inner section) out, creating more star formation within the arms of spiral galaxies—often in the globular clusters and open clusters there, described in Chapter 12, Star Formation, Star Development—than in the armless ellipticals.

M 104 is a spiral galaxy with a very large central bulge. The thin, spiral disk contains obscuring dust that makes this galaxy look something like its nickname implies, the Sombrero Galaxy. NASA

Quick Tips

1. The three main types of galaxies are spirals, ellipticals, and irregulars. The Milky Way is a spiral galaxy.
2. Galaxies also vary in types of luminosity, size, and degree of star formation.

Galaxies also vary a great deal in their level of activity. In the dimension of star formation, those the closest to a Fourth of July fireworks display are known as starburst galaxies. In these, stars are igniting at a rate up to 100 times faster than in our own galaxy. Star deaths are also common in these starbursts, happening as often as one supernova every 2 years. (For more information on supernova explosions, see Chapter 13, The Deaths of Stars.) Starburst galaxies are thought to resemble the conditions present in the early universe. Within their nuclei, some of these galaxies have quasars and some have miniquasars. These active galaxies produce their central power from phenomena that go beyond stars, as seen later in the chapter.

Milky Way Formation and Earth's Position

Our galaxy—long known as the Milky Way since a broad hazy swatch of it can be seen from Earth as a milky spill across the zenith—is about 13.7 billion years old. When it was young and not yet a mature spiral, it formed stars with alacrity in its halo and globular clusters. Gradually it collapsed to become flatter and with arms extending to create what looks like a spiral. Like other spiral galaxies, it is still forming stars, though at a much more modest rate, about seven new stars a year in the Milky Way. This formation is occurring in the gas and dust clouds of its disk and especially in the spiral arms. Our galaxy is growing much more by cannibalism, as is seen later in the chapter. (For more on the formation of stars, see Chapter 12, Star Formation, Star Development.) We are a typical spiral galaxy.

The Milky Way has four, possibly five, main spiral arms and several shorter spurs, each sweeping out tens of thousands of light-years from its "shoulders." But instead of twisting tighter as the galaxy orbits its center mass, these spiral arms remain in place—making them more like density ripples—while the stars themselves orbit the galactic center. Clouds of gas and dust become denser as they pass into a spiral arm, and this allows new stars to continue to form within them. The clouds in the arms are constantly being disturbed by inner gravitational attractions, gravitational disturbance from passing galaxies, shock waves from supernova explosions, ordinary galactic rotation, and nearby stars. All of these forces foment star development.

The Earth and the rest of our solar system lies in an outer area of one of the spiral spurs, about 28,000 light-years out from the galactic center, on a 15,000 light-year long "stub." The Sun will move out of this spiral arm in the distant future, since stars **out where the Sun is** orbit around the galactic center faster than the spiral arms do. (For more on the motions of planets and our solar system, see Chapter 3, Our Vantage Point on Earth.)

We live, then, in a modest exurb, but one in which a fair number of the houses are new, and more construction can occur at any time. Within about 33 light-years of our planet shine more than 300 stars, not at all a crowded condition like the bulge or nucleus of the galaxy but hardly empty either. Our area also lies inside one of the galaxy's "hot bubbles," relatively empty areas swept out by winds from a past supernova; these events were more common here not too long ago than they are now.

Quick Tip

The Earth lies on a short spur, an abbreviated form of a spiral arm, about two-thirds of the way out from the center of the Milky Way.

To see a distinct and impressive section of the Milky Way, go outside in summertime, as far from bright city lights and a bright Moon as possible. To the Northern Hemisphere eye, our galaxy stretches like a thick, blurry stripe across the zenith, nearly anchored on both horizons. This is actually an excellent view of one of our galaxy's spiral arms spread across the night sky. Southern Hemisphere viewers see a different section of our galaxy, not as long a band but still substantial.

The "milky" light is the combined light of many, very distant stars. Also, every star visible to the naked eye, both in and out of the "milk spill," is part of the Milky Way. The only other objects in the Northern Hemisphere **visible to the naked eye** sky are our sibling planets (one or two of them is usually visible at any given time) and the neighboring Andromeda galaxy. It is discussed later in the chapter along with galaxies visible only from the Southern Hemisphere.

The Disk, Bulge, and Black Hole of the Milky Way

The disk occupies a vast section of our galaxy. The spiral arms are part of it, but this feature slices through the entire galaxy, in its largest horizontal dimension. Its diameter stretches about 100,000 light-years. This disk is a flatter area of gas and dust clouds, home to both new and old stars. It has been differentiated into two overlapping disks, a thin disk and a thicker one. The thin disk, the flatter part of the disk system, is about 9 billion years old and 2000 light-years thick. Almost all of the stars near to us are in this thin disk, including the Sun; in the outer reaches of the galaxy, the youngest component of the thin disk is only about 400 light-years thick. The thick disk, 5000–10,000 light-years thick and extending above and below the thin one, contains more older stars than younger ones.

These disks are an orderly feature of the Milky Way galaxy, orbiting the galactic center steadily and in the same direction. But individual stars within the system can experience local motions that vary from this, bobbing up and down within the disk and even below or above it for stretches of time, as they respond to short-term gravitational tugs from nearby stars. The mass of the disk always pulls these stars back in, though a star-straying may last a few tens of millions of years. Should this happen to our Sun and solar system, the configurations of stars visible in our night sky would change, though extremely gradually.

Central Bulge

The central bulge of the Milky Way is bright with a concentration of billions of old stars, thick interstellar dust (which prevents astronomers from seeing most of the stars there as visible light), and other astronomical phenomena, especially within about 3000 light-years of the galaxy's central point or nucleus. The bulge is probably not exactly spherical, perhaps shaped more like a stubby cigar. The most outlying areas of it are up to 20,000 light-years away from its center point.

The density here is absolutely impressive and becomes even more so as one reaches the nucleus, which is only a few light-years across. Several millions stars, many supernova shells, twisted magnetic fields, black holes large and medium, and objects emitting strenuously in the radio and infrared radiations, all of them occupy this space. Even a population of young, hot stars shines here. In one particularly concentrated area, at the very galactic center, 3–4 million solar masses are crowded into an area not much bigger than our solar system, that concentration, notably, is a massive black hole. Its name is Sagittarius A*.

The central bulge and nucleus are home to the full array of stars in their later life-stages, as described in Chapter 13, The Deaths of Stars. There are close binaries (hundreds and hundreds of them), neutron stars, pulsars, white dwarfs, black dwarfs, brown dwarfs, unusual threads of hot gas, a ring of thick dark clouds where new stars are still being born, x-ray binaries, x-ray emissions as objects are pulled into the central black hole (which flickers as it eats), and, of course, masses of the thick dust that can be catapulted out as stars die in crowded conditions. In this geriatric concentration, the average star is 11–13 billion years old, and only a few stars are young, or even as young as our middle-aged Sun.

Black Hole

The role of black holes at galactic centers such as the Milky Way's extends even beyond its "star-eating" activity (it consumes about one whole star every few thousand years, as mentioned in Chapter 13, The Deaths of Stars). Black holes and their galaxies seem to evolve in tandem, for example, even though areas more than a light-year or so from the hole are the only ones that should really be affected by

NGC 4414 is a dusty spiral galaxy with a small central bulge. Our Galaxy, the Milky Way, would probably look something like this if we could travel outside and look back. NASA

it. The jets of material that black holes propel outward can both increase star formations and shove the gas and dust star nurseries completely away.

Black holes affect whole clusters of galaxies also. This has proved easier to study in neighboring galaxies (such as M87) than in our own, but conditions at the center of the Milky Way are likely to be similar. Jets of material here being are whipped outward by the twisted magnetic fields of the black hole, expelled at close to the speed of light from the galactic center. This force is actually enough to warm the spaces *between* galaxies, making them hotter than they otherwise would be. (The outflow of material in these jets also keeps supermassive black holes from becoming even more supermassive, though they certainly continue to grow.)

As the warm gas created by the jets in intergalactic space cools out there, that means it loses energy, which allows it to fall back in toward the galaxy. That infall next creates great cavities, empty areas between galaxies, from which perhaps a trillion solar masses have been swept clean. In the early universe, a process like this could have vastly increased star formation in a galaxy as the force of the jets on the way out disturbed the clouds of gas and dust. Astronomers think that they see evidence of disruptive jets emitted by the Milky Way's black hole as recently as 10,000 years ago.

It is, then, as though a galactic center's black hole is exhaling, then inhaling matter in cycles that last hundreds of millions of years. This makes black holes part of the ecology of the universe much more so that was originally suspected.

When galactic collisions occur, their black holes eventually merge. It is thought that, when the Milky Way merges with Andromeda, our combined black hole will reach the immense mass of about 100 million Suns.

The implications of black hole activity for the evolution of the universe have not yet been entirely discerned, but our ultimate fate clearly is affected by what goes on in between galaxies, and amid galactic clusters. (See Chapter 17, The Fate of the Universe, for more on this subject.) And the role of black holes has implications also for the early history of the universe: perhaps the black holes even evolved before most of the stars in their host galaxies were born, proceeding from the deaths of the very first generation of massive stars.

Milky Way's Interstellar Medium, Star Clusters, and Halo

The interstellar medium, the material among the stars, is of lower density than the stars, of course, but it is extremely important. As has been indicated earlier, this is the location of the gas and dust clouds, which function as star nurseries. The Milky Way is home to an abundance of cold atomic hydrogen gas, which stretches within the disk layer as far out as 80,000 light-years from the galactic center (well beyond the Sun's position). Dust, thick within the spiral arms, tends to exist closer to the center, though it extends at least as far out as the Sun's orbit. Large molecular clouds, rich in hydrogen (which formed molecules as the atomic hydrogen cooled in places) are also present along with other cloud molecules such as water, ammonia, and ethyl alcohol. These clouds add up to a mass in the Milky Way of about 5 billion solar masses. Clumping regularly to form stars, they have an important role in a middle aged, but still dynamic, galaxy.

Star Clusters

The Milky Way is also rich in specific star-forming clusters, including a globular cluster only 5 billion years old and one that may be a super star cluster similar to those common in the early days of the universe but only 3.5–5 million years old. These two clusters, in particular, form excellent evidence that stars are still being born in our galaxy. The first of these clusters is small with only about 2000 stars, while the super cluster has at least 100,000 stars packed into an area only about 6 light-years across. Almost all of the globular clusters found in the Milky Way are as old as the galaxy itself (13.7 billion years). In total, the Milky Way seems to contain 150 globular clusters, most quite ancient and all in the halo of the galaxy. (Andromeda has about 500 and giant elliptical galaxies have thousands.)

Halo

Quite diffused, the halo of the Milky Way forms a kind of sphere around the galactic center and bulge. In it live a smattering of mostly individual stars, generally old and dim, along with relatively little gas, and cold blobs of hydrogen gas that do not rotate with the disk at all. The smaller numbers of younger stars here have probably been pulled away from the disk or even "stolen" from other galaxies that passed by. Most of the halo, though, dates back to 13.7 billion years ago, the time of our galaxy's birth. It extends outwards about 50,000 light-years from the galactic center in all directions. Beyond the zone of starlight, at 15,000 light-years beyond the last star, lies a far outer disk of hydrogen, tilted around the galaxy like a jaunty hat. (Beyond even that) **Throughout our Galaxy** lies dark matter, whose nature and function is the subject of Chapter 15, The Role of Dark Matter.

The stars in the halo do not behave the way stars in the galactic disk, such as our Sun, do. They orbit the galactic center in random directions. Moving every which way, they sweep above and below the disk, too, careening through it so fast that the gravity of the disk barely slows them down. We can see one of these erratic halo stars almost any clear night and from both hemispheres: Arcturus, the fourth brightest star in the Earth's sky, (is) **may be** a halo star, on its way through our area of the disk.

Though the halo sounds like a swarm, it is not a dense one. There are actually not many stars out here at all, relatively speaking. A great deal of the mass of the halo—probably 90 percent of it—is some form of the dark matter mentioned above and to be discussed in the next chapter.

Ecology and Evolution of the Milky Way

Our galaxy, now 100 billion stars strong, formed originally from several proto-galactic clouds which merged to become a single such cloud; the process was analogous to the protostar stage of star formation discussed in Chapter 12, Star Formation, Star Development. This spherical, diffused cloud gradually formed stars and, over a few hundred million years, collapsed to become a disk rotating around the galactic center. Our black hole was probably formed at some early, point. As the disk swirled and, as the first stars died, more stars continued to be born.

But all is not this straightforward, as astronomers have been recently learning. They have found some unusual "corners" of the Milky Way. For one thing, some of the globular clusters appear to have come from elsewhere, since they are the wrong age to have originally been part of our galaxy. There are also groups of stars, often formed into "star streams." Several of these galactic features have been recently classified as the leftover bodies of an old, "swallowed-up" galaxy, now stripped of most of its gas and structure. These star streams move together within the Milky Way but in a different way from their stellar "neighbors." Dwarf galaxies exist just beyond the Milky Way, too, with three such "satellites" recently found at 100,000 light-years, 200,000 light-years, and 640,000 light-years from the Earth. Those in our immediate galactic neighborhood, and these appear to have become stripped of both stars and interstellar material as they passed through the edges of our galaxy and continue to do so. One of these small, very near neighbors—made of ordinary, visible matter—is down to only about 150,000 stars. All of these anomalies indicate collisions, capture, and cannibalism at the intergalactic level. The Milky Way has been partly assembled, then, by material from elsewhere, pulled in by our gravitational strength.

Gravity can work the other way too, of course. Astronomers think they see disruptions in an area of the Milky Way's disk, where stars have been shoved closer together, after one of the larger dwarf galaxies passed very close to us. Though small, it had a local effect.

Our galaxy is always engaged in basic recycling, too. In this process, called chemical enrichment, stars forge new, heavier, elements, propelling them out into space in substantial amounts via both their ordinary, streaming, stellar wind and in the shock waves of the much more energetic supernova explosions when they occur. The chemical elements are then taken up by the next generation of stars, born in the interstellar gas and dust clouds; these clouds had snared the material as it hurtled by at close to the speed of light, then cooled down over millions of years to star-forming temperature. If these chemical elements were not initially trapped in this way, in hot bubbles which can expand from 10 light-years across up to 1000 light-years across, the supernova material would accelerate entirely out of the galaxy. In fact, sometimes it does: if ejected star material does not intersect enough of the interstellar medium, it races out well into the galactic halo in a kind of "volcanic" eruption.

Galactic change occurs at even larger scales also, as is mentioned later in the chapter.

These colliding galaxies have been nicknamed "The Mice" because of the long tails of stars and gas emanating from each. The pair will eventually merge into a single larger galaxy. NASA

Quick Tips

1. The basic elements of the Milky Way galaxy are its disks, its central bulge, and its supermassive black hole at the nucleus, emitting jets of material just before it crosses the event horizon, its interstellar medium, its star clusters, and its halo.

2. Our galaxy is always changing, through the chemical enrichment processes that stars conduct, the galactic assimilation of hydrogen gas masses, and cannibalism at an even more dramatic level.

Andromeda Galaxy

Our closest galactic neighbor, Andromeda, lies 2.3 million light-years away. Another spiral galaxy, it is only slightly larger than the Milky Way. Andromeda is, in a sense, easier to study than the galaxy in which we are embedded, and much has been learned about spirals in this way. Andromeda, also known as M31, is now known to have:

- A faint disk within its halo, extending to about 230,000 light-years from its center. This suggests a mostly "digested" portion of one of the nearby, smaller, galaxies in our Local Group.

- Many globular clusters, one as young as 2 billion years old. This cluster may, in fact, be an entire dwarf galaxy captured by Andromeda then shredded a bit in the event.

- Swarming clouds of hydrogen gas snared by Andromeda as they passed by and now remnants of its own star formation.

- A stream of visible material flowing into Andromeda directly from the small dwarf galaxy NGC 205.

Far from city lights and a bright Moon, it is easy to see the Andromeda in the summer sky of the Northern Hemisphere. Look for it as a delicate blur near the Pleiades constellation, high in the eastern sky. This light left Andromeda 2.3 million years ago. It is possible to see Andromeda from the Southern Hemisphere also, though this is more difficult since it is so close to the horizon.

Our Local Group and Local Cluster

Galactic cannibalism occurs because there are other galaxies conveniently close to us and to Andromeda. Not every galaxy in the universe is in a Local Group, but we are. Members of our Local Group, anchored by the two largest (Andromeda and Milky Way) include:

- The Large Magellanic Cloud, the largest of our satellite galaxies and, like its smaller companion, a few billion stars strong. It lies about 160,000 light-years away.

- The Small Magellanic Cloud, a dwarf already stretched and pulled to the point where its outer edges are up to 250,000 light-years away from us, though its inner edge is much closer.

- M32, NGC205, and M33 (the Triangulum galaxy).

- Sagittarius, a dwarf elliptical and, at about 50,000 light-years away, the closest to us of all. It is still in the process of being torn apart by the Milky Way, and this should require about 100 million years more to complete.

- Leo 1, another dwarf elliptical.

- Ursa Major dwarf, at 350,000 light-years away, has only about 40,000 stars left. This galaxy may be more significant than that sounds, however, as is mentioned in Chapter 15, The Role of Dark Matter.

- Many more dwarfs and surely some yet to be discovered. Two—Bootes and Canes Venatici—were discovered just in the summer of 2006. It is possible that we have up to 100 small, satellite galaxies.

Both the Large and Small Magellanic Clouds can be seen with the naked eye from the Southern Hemisphere (as can the galaxy M83, the Milky Way, and, with difficulty, Andromeda). They are visible in summer (November–February) high in the southeastern sky. The Large Cloud, home to 20 billion stars, most of them old, is also actively engaged in star formation. A nebula within it, 30 Doradus, is composed of the new, blue stars less than 3–5 million years old. But as a dwarf galaxy and satellite to the Milky Way, the Large Magellanic Cloud will not last forever. Observations in the radio part of the spectrum show dark threads of gas being pulled out from it continuously by the gravitational force of the Milky Way.

These dwarf galaxies of our Local Group were probably formed early in the history of the universe. They now function, however, as more like the raw material for the aggrandizement of their larger companions.

Since our smaller companion galaxies orbit us, complex patterns of gravitational interactions often pull them close enough to be regularly "raided" of material, from clouds to stars. Some of this can be seen now as "tails" tugged from the near side of the two Magellanic Clouds. These two satellite galaxies will eventually be entirely assimilated into the Milky Way.

For once, then, we are above average: on an average-sized rocky planet orbiting an average Sun, in a modest part of our galaxy, our Milky Way is at least, along with Andromeda, the major player in this Local Group of galaxies. We two "big shots" orbit each other, too, every 10 billion years or so, gradually approaching like a shy but intrigued pair of partygoers.

Our Local Group is part of an even larger aggregation of galaxies called the Virgo Cluster. It is 10 million light-years across and contains a mass equal to about 100 trillion Suns. We are all held together by dark matter (the subject of Chapter 15, The Role of Dark Matter). Clusters like ours are still forming in the universe, though the process was at its peak about 5 billion years ago. The clustering going on now is more often mopping up the dwarfs and playing "bumper cars" with the larger galaxies within the cluster.

The Virgo Cluster contains all of our Local Group of galaxies plus many thousands of other galaxies. All are in complex motions with regard to each other, but our Local Group is being tugged, overall, in the direction of the center of mass of this cluster. We are still able to move away from it, but at a speed 248 miles (400 kilometers) per second slower than we would be if it were not attracting us. Most of the other local groups within the cluster are increasing the spaces between them and the Virgo Cluster's center of gravity also.

The immense galaxies toward the center of our Virgo Cluster, now ellipticals and already swollen (and lumpy) **(delete)** from previous collisions, lie in wait there. They are ready to cannibalize us should the future of the universe lead toward inward collapse. (This seems unlikely, at least.)

Quick Tips

1. Andromeda galaxy is our closest companion, another spiral, and a bit larger.

2. Our Local Group of galaxies is persistently engaged in cannibalism, with Andromeda and the Milky Way assimilating a great deal of material from our smaller companions who number more than a dozen.

3. Our Local Group is also part of a cluster of local groups, called the Virgo Cluster. This larger aggregation is composed of more than 2000 galaxies.

Local Supercluster, Great Attractor

Our Local Group and the larger Virgo Cluster are also within an even larger aggregation of galaxies. Spread over 150 million light-years, this is our Local Supercluster. It is composed of 2000–3000 galaxies.

The entire Local Supercluster is moving, too, toward an even more immense concentration of mass called the Great Attractor. The latter's mass is equivalent to that of tens of thousands of galaxies, which is what provides the powerful attraction. Whether our trajectory toward it will be maintained is discussed in Chapter 17, The Future of the Universe.

Quasars

It is time to leave the Milky Way behind, abandoning the somewhat parochial perspective of the last several sections. Our galaxy does not contain a quasar at its nucleus. Only one out of every 10 million normal spiral galaxies does. Yet these quasars, the most powerful objects in the universe, have an immense influence upon the evolution of galaxies. Not enough is known about them to fully limn this process, though more is being discovered all the time.

Quasars can look like stars if seen from far away (and most of them are very far away). The size of each of them is about that of our entire solar system. They sit at the center of a spiral or an elliptical galaxy, radiating energy in most or all wavelengths. Because of the current expansion of the universe, distant quasars can be speeding away from us at about 94 percent of the speed of light. Even stranger, the brightest quasars shine brighter than 1000 entire Milky Way galaxies, and even typical ones are about half that bright. These odd objects can vary their brightness, too, even by 30 percent, over just a few weeks. Their power is more efficient, more extreme, even than the nuclear fusion that powers stars in a quasar, 10 percent of the mass can be transformed into energy. (Stars like the Sun manage about a 1 percent conversion rate.)

What are these objects? The clue seems to lie in the fact that most are very far away. The most distant quasar discovered so far was born only about 870 million years after the Big Bang, and so its light has taken most of the lifespan of the universe to reach us. Our current instruments cannot actually see back to the time of the first stars.

The scene for quasar development was set because early galaxies are likely to have contained substantial black holes. These black holes sopped up massive amounts of the excess gas and dust clouds after the early and vigorous bouts of star formation in the early universe. Next, they consumed some of their nearby stars. There was more loose fuel available in galaxies in those days than there is now, and there were also more galactic collisions since the galaxies were closer together. The black holes grew huge on all this "nutrition," some of them reaching the mass of a super-massive black hole, between 1–5 billion solar masses. (This is far heftier than the Milky Way's present black hole.) The earliest such objects date back to when the universe was only about 600 million years old. A black hole this massive develops a broad accretion disk, with matter swirling chaotically in a disk around the hole as it enters the event horizon. The gases become extremely hot, and forces that are not well understood launch massive jets of material perpendicular to the disk. It is the energy released by the in-falling matter that produces the quasar.

The more common existence of supermassive black holes in the very early universe tells us that black holes did indeed develop before galaxies had completed their formation. Ordinary black holes today can only become supermassive black holes and then engender quasars if their fuel supply somehow increases. These days, that seems to happen only in the collisions of two large galaxies. When Andromeda and the Milky Way collide, for example, gas would stream down to their merging black holes, perhaps creating a local quasar, but only then.

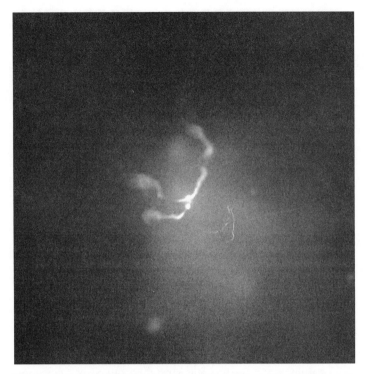

The radio galaxy 3C 75 actually contains two galaxy nuclei, each shooting out double lobed jets into the intergalactic medium. Each active nuclei contains a supermassive black hole accreting gas and producing the jets. This image is a composite of radio observations, which see the nuclei and jets and an X-Ray image, which sees the glow of super-hot gas surrounding the entire system. NASA

What would a galactic collision on this scale mean for our small, but precious planet? Probably nothing. First of all, the Sun is likely to have died by then anyway, eliminating life from Earth. Even if a collision were to begin relatively soon, it could take hundreds of millions of years for disruptions to become detectable. And for two galactic centers to approach each other closely, forming an immense elliptical galaxy out of two spirals—and possibly firing up a quasar—would require millions of years more than that. Even after that point, if the Earth had somehow survived and was still in an outlying area the way we are now, what would change mostly is the look of the night sky: it would be brighter because of all the new stars nearby, formed in the collision.

Interactions of Galaxies in Spacetime

As we have seen, galaxies build up by merging material from elsewhere as well as by creating new stars. This cannibalism has been going on since the universe was 1–6 billion years old. Recently, the largest structure ever discovered in the universe was measured as 200 million light-years long, composed of galaxies and immense pockets of gas which date back to about 2 billion years after the Big Bang.

Steadily, galaxies, local groups, clusters, and superclusters have grown. Space between superclusters has stretched out also, since mass has become concentrated and the universe is expanding. Some areas of the universe have even become almost empty in between, and these are called void. **(you need the "s," its voids)** Though local groups are quite bound together gravitationally, local clusters do not exhibit as much such "glue." Even so, our supercluster is able to proceed toward the Great Attractor.

Where will it all end? Though the universe appears to be expanding now, even if not uniformly, it is not completely clear whether this will continue. How well will the gravitational "glue" work? Will it be overwhelmed by other forces? And why is this especially hard to analyze? Exploration of these questions requires a look at the mysterious dark matter and dark energy of the universe, in the next two chapters, and then at the puzzle of the expansion and contraction in the future of the universe.

The future for some galaxies looks quite grim. The dwarfs in our Local Group are being gradually dismembered. And, much farther away, another phenomenon is occurring. The galaxy C153, for example, has had all but one of its spiral arms ripped right off. Once an elegant spiral about a billion light years from us, its "mistake" was to be pulled into a cluster of galaxies called Abell 2125.

Problems began even before C153's gas clouds and stars got near enough to be stripped away by hot gas in the cluster. At high speed, the galaxy hit a vast nimbus of gas encircling the Abell cluster which subjected it to "ram pressure stripping." Its own gas clouds were blown outward, gone forever, though its stars remained. Along the way, this disruption caused a brief, local flurry of star formation in C153, though this was like a mere whistle in the dark. Few of those stars should still be in place 100 million years from now, and new ones will not be forming either, since the gas clouds are gone. Astronomers seeking reasons for the fact that spiral galaxies are less common in galactic clusters today than they once were need to look no farther than C153.

In this chapter you have learned that:

 Spirals, ellipticals, and irregulars are the main types of galaxies, with the majority of the large galaxies (including the Milky Way) being spirals.

 Ellipticals, especially dwarf ellipticals, are often, though not always, the next stage after spirals have partially collided, and large or giant ellipticals can be the result of a complete merger of two spiral galaxies.

 Galaxies vary greatly in size, luminosity, degree of star formation, and level of activity in their nuclei.

 In our Milky Way galaxy, the Earth and the rest of our solar system are positioned on one of the spiral spurs and in a "hot bubble" area where supernova explosions were common in the relatively recent past but no longer are so.

 Our immediate area is also part of the galaxy's thin disk.

Star formation in galaxies occurs most often on the spiral arms and spurs or in the globular star clusters of the halo.

 The other typical parts of a spiral galaxy such as the Milky Way are a central bulge, a black hole at the nucleus (which emits jets), an interstellar medium, and dark matter.

Galaxies evolve through interactions among themselves, through chemical enrichment as stars are born and die, and through the actions of their black holes.

✔ Our galaxy, and many others, are part of a local group, a local cluster, and a local supercluster.

✔ Quasars, rare now, are the most powerful objects in the universe and are powered by supermassive black holes as they consume an exceptional amount of matter.

Test Yourself

True/False Questions

1. The average galaxy contains about 1 trillion stars.
2. Large elliptical galaxies are the most common type.
3. Irregular galaxies were more common in the early universe.
4. The Milky Way has three spiral arms.
5. Most star formation in our galaxy is occurring in the central bulge.
6. The main activity of a black hole is the consumption of stars.
7. Galaxies typically contain clouds of atomic hydrogen as well as clouds of molecular hydrogen.
8. Globular clusters are usually positioned in the halos of galaxies.

Completion Questions

1. The three main types of galaxies are _____, _____, and _____.
2. The galaxies within which star formation is occurring 100 times faster than in the Milky Way today are called _____ galaxies.
3. The Milky Way formed by about _____ years ago.
4. The "neighborhood" of the Milky Way is a spiral spur about _____ light-years long.
5. The galaxies visible to the naked eye from the Northern Hemisphere are _____ and _____.
6. The most orderly part of a spiral galaxy is its _____.
7. The area of a spiral galaxy which is the densest in stars is the _____.
8. The galaxy recycles star material in a process called _____.
9. Andromeda galaxy is _____ light-years from the Milky Way.

Multiple Choice Questions

1. The oldest stars in a galaxy are typically in
   a. The central bulge
   b. High-collision areas
   c. The interstellar medium

2. The Earth's position lies how far out from the galactic center?
    a. 15,000 light-years
    b. Considerably farther than the Sun's
    c. 28,000 light-years

3. Black holes are studied primarily
    a. By their gravitational lensing
    b. In their quasar dimension
    c. Through their x-ray radiation

4. The halo of our galaxy extends about how far out from the galactic center?
    a. 50,000 light-years
    b. 28,000 light-years
    c. 15,000 light-years

5. The number of galaxies in our Local Group is about
    a. 5–7
    b. 13–20
    c. 40–50

6. The Magellanic Clouds are
    a. Star formation areas
    b. In the halo of our galaxy
    c. Dwarf galaxies

7. As part of the Virgo Cluster, we are moving toward
    a. Sagittarius A
    b. The Great Attractor
    c. Our galaxy's black hole

ANSWER KEY

True/False
1. F 2. F 3. T 4. F 5. F 6. F 7. T 8. T

Completion Questions
1. Spirals, ellipticals, irregulars 2. Starbursts 3. 13 billion 4. 15,000
5. Andromeda and the Milky Way 6. disk 7. Central bulge 8. Chemical
enrichment 9. 2.3 million

Multiple Choice Questions
1. a 2. c 3. c 4. a 5. b 6. c 7. b

PART V

CHAPTER 15

The Role of Dark Matter

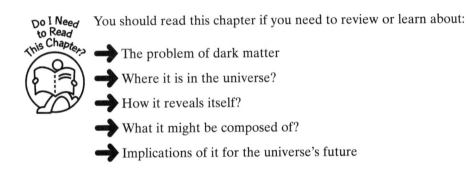

You should read this chapter if you need to review or learn about:

➡ The problem of dark matter

➡ Where it is in the universe?

➡ How it reveals itself?

➡ What it might be composed of?

➡ Implications of it for the universe's future

Get Started

Dark matter, which emits no radiation at all, is the name for an unknown sub-stance that supplies the gravity needed to hold galaxies and the universe together. Most of the universe is actually composed of it and of dark energy (the subject of the next chapter), while visible, ordinary matter comprises only 5 percent of the universe. Dark matter was not actually detected until late, 2006, and remains a mysterious substance.

The Problem of Dark Matter

Astronomers believe that the universe is composed of 4 percent ordinary matter, 20 percent dark matter, and 76 percent dark energy. Only once has anyone directly observed dark energy and neither it nor dark energy has been fully analyzed. The problem of dark energy will be addressed in the next chapter and that of dark matter in this one.

Dark matter is known to exist because, without it, there would not be enough gravity to hold the galaxies and clusters of galaxies together. Outer stars would simply escape. Since we can discern the luminosity of a galaxy, and usually figure out its rotation, astronomers can quite clearly estimate its mass by asking the question: how many stars are needed to add up to the observed luminosity? The mass we see as ordinary matter is not enough to provide the necessary gravity. There must be a lot more "stuff" out there. This is true throughout the galaxy but especially so in its outer reaches, which would fray constantly without the existence of dark matter's extra mass.

And the universe is indeed held together. Chapter 14, Galaxy Formation and Development, described the aggregations of ordinary matter not only in galaxies but also in local groups of galaxies, in clusters of local groups, and in superclusters of galaxies. The largest grouping of galaxies actually observed so far is a line of galaxies about 200 million light-years long. Assemblages like these require a lot of "glue." The invisible part of this gravitational power is called dark matter. Astronomers call it dark matter because it produces no starlight yet exerts a gravitational force.

In current theories, dark matter was the first thing to coalesce in the universe, and it was its gravity that allowed the galaxies themselves to form. The first differences in density within the quark-gluon plasma that formed after the Big Bang would have been found in it. Later the cohesion of the protogalactic

The cluster of galaxies EMSS 1358+6245 is detected in this Chandra X-Ray Observatory image as a large volume of super-hot gas. What holds this gas together? There is not enough gravity from all of the stars and galaxies in the cluster, so astronomers conclude there must be another type of matter, so-called dark matter, that provides the necessary gravity. NASA

clouds from which the galaxies formed would have required earlier clumping of this dark matter. Dark matter currently contributes more than 80 percent of the gravitational force of the universe.

Dark matter is not dangerous—it bears no resemblance to antimatter.

When a matter and an antimatter particle (e.g., a proton and an antiproton) collide, they annihilate each other instantly in a burst of energy. When two dark matter particles encounter each other, however, they usually pass right by each other. And when a dark matter particle encounters an ordinary matter particle, they just give each other a gravitational tug.

Where Is It in the Universe?

Dark matter lies primarily in vast halos around galaxies, forming about 90 percent of the material in those locations. The dark matter halo of the Milky Way, for example, is about 300,000 light-years in diameter. It also forms a nimbus around the larger aggregations of galaxies such as the clusters and superclusters. And dark matter exists in lesser amounts within galaxies—there is very little within our solar system, for example, though significant amounts of it are present within a few hundred light-years from here.

Dark matter dominates galactic clusters. It is required in order to explain the velocities within a cluster and to provide the gravity to power the x-ray gas seen within clusters.

This strange material is not uniformly distributed in clusters the way it is in individual galaxies, however. A few whole galaxies have been detected where dark matter is almost all that is there. One of the Milky Way's dwarf galaxy companions, the tiny Ursa Major (mentioned in Chapter 14, Galaxy Formation and Development), may be almost entirely dark matter. And another galaxy, VIRGO HI21, may be yet more completely dark, its ordinary matter component is just a tenth of one percent of its mass.

Chandra finds massive X-Ray halos in most every galaxy cluster. This indicates that Dark Matter is everywhere in the Universe. NASA

Does dark matter really exist? A substance so mysterious may sound actually impossible, particularly if one remembers some of the other odd constructs in the development of science, from the medieval aether and phlogiston to the Elizabethan humors, and beyond. And indeed dark matter presents some challenging puzzles: Why does it not form a halo around every galaxy? Why is there a lot more of it around some galaxies than others? Does the amount of it change as a galaxy evolves? as a group, cluster, and supercluster evolves? as the universe ages? Why aren't there even more dwarf galaxies in local groups made predominantly of it?

Astronomers have detected it so far only in a collision between two **clusters of galaxies**, where it behaved as though it was relatively unaffected by the merger. Ordinary matter was slowed down in the "crash."

How Does It Reveals Itself?

As has been mentioned, dark matter usually signals its existence because it is what is required to make galaxies and larger aggregations of them hold together, move as they indeed move, and exhibit the hot x-ray gas profiles among them that they do. It reveals itself yet more clearly through an effect called gravitational lensing. In this situation, the mass of an object can bend light—and this can be the light from a distant galaxy on its way toward Earth, for example. If the mass is large enough and if the objects are lined up correctly from our point of view, the bent light can create double and quadruple images of faraway objects. The unseen mass of dark matter is indeed creating this effect here and there. And astronomers can tell how much of it is doing so out there.

Dark Matter's Composition

What is it? Explanations so far are only partial and begin with what dark matter is not:

- Dark matter is not clouds of dust. If it were, there would need to be so much dust that we could hardly see the visible light of any stars.

- Dark matter is not hydrogen in any form because that chemical reveals itself as it alters the starlight, or changes the spectrum, of stars farther away.

- Dark matter is not extra black holes of the ordinary active sort, since they indicate their existence by x-rays as they consume material.

Instead:

- Some of the dark matter might be dim objects—such as planets, brown dwarfs, and various kinds of dead stars (including black holes in quiescent stages)—all made of ordinary matter but not emitting enough radiation to be detectable. In our own galaxy's halo, however, it seems that not more than 50 percent of the dark matter could be of this type, and, within clusters of galaxies, these types of bodies could account for much less.

- Or, most likely, dark matter could be exotic subatomic particles. These particles must contain mass but are capable of only weak interactions with other particles and each other. (Otherwise they would constantly bump into normal matter, revealing their presence.) Physicists call these "weakly interacting massive particles" (WIMPs). Their exact nature remains a mystery.

Implications

Dark matter is extremely important for the analysis of the future of the universe. This is because it provides gravity, and gravitational power is what determines whether the universe expands infinitely or collapses inward—or oscillates back and forth in some fashion—in tether to its density of ordinary matter and dark matter combined. This subject is pursued further in Chapter 17, The Future of the Universe, after a consideration of the problem of dark energy in the next chapter.

It's a Wrap

In this chapter you have learned that:

 Dark matter is far more common in the universe than ordinary visible matter.

 It must exist in some form or there would not be enough gravitational force to hold galaxies and other aggregations together.

 It reveals itself, to the extent that it does, by acting as gravity and by being relatively unaffected by a galactic collision.

 It varies in its concentration in space and, to some extent, throughout the history of the universe.

Test Yourself

True/False Questions

1. Dark matter forms a halo around every galaxy.
2. Dark matter interacts with ordinary matter through gravity.
3. In the early universe, dark matter probably coalesced after ordinary material.

Completion Questions

1. The largest aggregation of galaxies discovered so far is _____ light-years long.

2. Up to _____ percent of dark matter could be ordinary matter that is too dim for us to detect.

Multiple Choice Questions

1. Galaxies made primarily of dark matter are
   a. The Ursa Major dwarf
   b. VIRGO HI21
   c. Both

2. It is possible that some of the dark matter may be
   a. Subatomic particles
   b. Small asteroids in our solar system
   c. Interstellar dust clouds

ANSWER KEY

True/False
1. F   2. T   3. F

Completion Questions
1. 200 million light-years   2. 50 percent

Multiple Choice Questions
1. c   2. a

CHAPTER 16

The Role of Dark Energy

Do I Need to Read This Chapter? You should read this chapter if you need to review or learn about:

➡ The problem of dark energy

➡ How it is detected?

➡ Whether it is changing?

➡ What it comes from?

Problem of Dark Energy

About 76 percent of the universe seems to be dark energy. As with dark matter, it is invisible. And it seems to be quite uniform, not gathered into galaxies (primarily their halos and other aggregations) the way as dark matter is. It could be compared to the silences in an important conversation; they can be as significant as the words spoken.

Dark energy will never be directly visible, since it is akin to a force field, consistently pushing the intergalactic universe outward at its largest scale—that of superclusters and beyond—at an accelerating rate. At the smaller scales of galaxies, local groups, and clusters, gravity's inward pull is stronger than dark energy's outward shove.

Until fairly recently, astronomers expected the opposite of this, conjecturing that the steadily aggregating masses in the universe—the local groups joining clusters, the clusters steadily joining to make superclusters, and so on—would continue to slow down the expansion of the universe, which had blasted outward in the Big Bang, an energy echo of which is still (barely) detectable today in the cosmic microwave background of the universe. (See Chapter 1, The Universe in Time and Space.) Dark energy is a line of research only about 10 years old, with much more attention needed.

Dark energy is an extremely important mystery since this is the very force likely to determine the future of the universe: will it ultimately collapse? expand forever? remain in a steady state? or undergo an accelerating expansion?

One way to think of dark energy and gravity as opposing forces affecting the future is to imagine tossing a ball up into the air, four different times. In the first toss, the ball rises up from the surface of the Earth, but since the Earth's gravity is always pulling back, it eventually slows down and falls back down to Earth. This is a "closed universe," one that will eventually collapse. We do not appear to live in this kind of universe, however. So it is time for the second toss.

Here, you throw the baseball up so fast that it completely escapes the pull of Earth's gravity. It flies off into space, always moving away, forever. This is like the scenario of an "open universe" that will expand forever, only slowing down slightly. We do not appear to live in this kind of universe either.

In a third toss, you throw the ball up at exactly the *escape velocity* of the Earth, just enough to allow it to overcome the power of Earth's gravity (spaceships have this velocity). The ball will always be moving away from the Earth, though just barely, and it will always be slowing down. This is the analogue to a "flat universe" filled with matter. Until recently, we thought that this was the kind of universe we live in. Not so.

Recent observations show that the universe is accelerating, not slowing down, and requires some sort of dark energy to do so. Hence, this is what happens in a fourth toss: you throw the baseball, it travels up, then slows down, somewhat, because of Earth's gravity. But, as the ball gets farther away, the gravitational pull of the Earth gets weaker. A small negative gravity, or dark energy—too weak compared to gravity when the ball was close to the Earth—now starts to determine the trajectory of the baseball. As the ball travels farther and farther away from the gravitational tug of the Earth, this force begins to dominate gravity. The ball then starts to accelerate away from the Earth—at an ever-increasing velocity. This appears to be the kind of universe we live in.

How Is It Detected?

If the expansion of the universe is indeed accelerating because of dark energy's outward force, that expansion and pattern of the universe's mass should somehow reveal its existence. And it does, in primarily five ways. First, very distant white dwarf supernovas (called Type 1a) are farther away than they should be, given their brightness; this can be surmised because these objects do have an intrinsic brightness, which can be compared to their apparent brightness (and their spectra, or light profiles, then tells astronomers their distance from us). Second, the specific distribution and distance of these supernovas does indeed fit with a universe of 70 percent dark energy. Third, the distribution of several millions of the galaxies in the universe and small fluctuations of temperature in the cosmic microwave background (which reflects that galactic distribution) indicates the kind of universe that requires dark energy in order to exist. Fourth, the distances to 26 different clusters of galaxies, as discerned by studying the hot x-ray gas among them, also fit the same picture. And the newest detection method is to study gamma ray bursts to search for dark energy. Though these lines of evidence may sound obscure and preliminary, astronomers consider them very persuasive of the existence of dark energy. Galaxies do indeed appear to be careening away from each other much more rapidly than they were 5–6 billion years ago.

Because supernovae are such powerful explosions, we can see them at great distances. Here a supernova (the bright star in the lower left) has exploded in the galaxy NGC 4526, rivaling the light from all of the stars in that galaxy combined for a few brief weeks. NASA

Difference: 1997-1995

This fuzzy image taken by the Hubble Space Telescope shows a field of extremely distant galaxies. When images taken in two different years were differenced, one of the most distant supernova ever found was revealed. By studying distant supernovae, astronomers can measure the expansion and acceleration of the Universe. NASA

Is dark energy another form of gravity which we do not understand? Gravity is thought of as an attractive force, pulling inward. Dark energy is a repulsive force, pushing outward. Yet they might somehow be related. Is dark energy a phase, an echo, of the "inflation" that originally engineered the dramatic expansion of the universe shortly after the Big Bang? (See Chapter 1, The Universe in Time and Space, for a description of this inflation.)

Changes in Dark Energy

Dark energy appears to be constant in the universe, though it may have been different very early in the Big Bang period. It seems to be uniformly distributed, as strong a force in one part of space as it is in another. It exists in every cubic inch of space.

In the early eras of the universe, however, matter was much more densely packed than it is today. Gravity was a stronger force than dark energy. By the time the universe had spread out and formed itself into many galaxies and other large aggregations, dark energy became stronger relative to gravity. This transition seems to have occurred about 5–6 billion years ago.

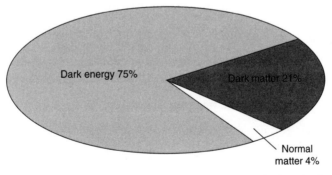

The matter you and the stars and planets are made of, what physicists call baryonic, or 'normal' matter makes up a mere 4% of the mass-energy density of the universe. Dark matter, which exerts a gravitational tug just like normal matter, makes up another 21%. Most of the mass-energy in the universe is in a form we are just beginning to understand called dark energy. NASA

In a dark-energy universe with the power of today's, the universe's 100 billion galaxies will gradually race outward, moving beyond the view of all our instruments as the universe expands. Our own galaxy will start to move away from the

Virgo Cluster, though over a long period of time. But, within our solar system and other small distances in space, gravity will be in control, holding us together.

What Does It Comes From?

Astronomers are currently exploring "vacuum energy" as probably the best way to describe the nature of dark energy. This is a quantum mechanical effect of spacetime in which particles of matter and anti-matter continually flicker in and out of existence, annihilating each other and thereby giving "zero point" energy to empty space. This would explain a force that seems to exist everywhere throughout "empty" space. (Space is not truly empty anywhere, even in the voids between galactic superclusters.) To some astronomers, this vacuum energy is unvarying throughout space, while others think it may be of a different strength in the very farthest reaches of space. Mathematically, it is similar to Einstein's cosmological constant. (See Chapter 9, Origins of the Puzzles of Spacetime and Gravity for more on Einstein.)

Other astronomers approach dark energy completely differently. They think it may not exist at all—and that we simply do not yet understand gravity itself in enough detail. This remains a possibility.

In this chapter you have learned that:

 Dark energy is accelerating the expansion of the universe.

✔ It may have changed throughout the history of the universe but is uniform throughout space now and in any one era of the past.

✔ It acts as the opposite of gravity.

✔ It is like a force field, probably based somehow upon quantum mechanical effects.

True/False Questions

1. Strong evidence for the existence of dark energy comes from supernovas.
2. About 25 percent of the universe is dark energy.

Completion Questions

1. Dark energy became a stronger force than gravity about _____ billion years ago.

Multiple Choice Questions

1. Dark energy is most significant at the scale of
   a. Galaxies
   b. Galaxies, groups of galaxies, and clusters of galaxies
   c. Superclusters of galaxies

2. Dark energy
   a. Is made primarily of dark matter
   b. Aacts oppositely from gravity
   c. Comes from the cosmic microwave background

ANSWER KEY

True/False

1. T   2. F

Completion Questions

1. 5–6

Multiple Choice Questions

1. c   2. b

◆◆◆◆◆◆◆◆◆◆◆◆◆◆◆◆◆◆◆◆◆◆◆◆◆◆◆◆

CHAPTER 17

The Future of the Universe

◆◆◆◆◆◆◆◆◆◆◆◆◆◆◆◆◆◆◆◆◆◆◆◆◆◆◆◆

Do I Need to Read This Chapter?

You should read this chapter if you need to review or learn about:

➤ Whether the universe will next undergo accelerating expansion, expansion, recollapse, or remain in a steady state

Get Started

To predict the universe's ultimate scenario, astronomers need to know the distance of the most distant objects that we can detect within the universe; whether there is more mass beyond that and how it is distributed; the mass of dark matter; and whether gravity or dark energy will become the predominant force of the future. Since only the first of these factors is indeed known in any detail, several different scenarios have been suggested as possibilities for our ultimate fate, though one is much more likely.

Acceleration Expansion?

It is probable that the universe will continue to expand at an accelerating rate. As discussed in Chapter 16, The Role of Dark Energy, this is the process in operation now—and it has been for the last 5–6 billion years. (That is, however, less than half the age of the universe so far.) It is not the case that galaxies (or any other aggregation) are rushing farther out into an existing empty space— the universe, spacetime itself is stretching, like a giant balloon moving outward in all directions, though in the case of the universe, there is no skin, no edge, only horizons beyond which we cannot see. According to our current understanding, this accelerating expansion will never slow down.

Galactic distances from us are determined by looking at their *redshifts*. The more distant a galaxy, the more its spectrum (the chemical profile of its light) is shifted toward the longer, redder wavelengths, the infrared. As space stretches, the length of the waves also stretches.

The most surprising aspect of this redshifting, and one that has been known to some degree for nearly 100 years, is that the farther away the galaxy is, the faster it is receding from us. The new part is that this expansion is now known to be accelerating; the faster rates, for the most distant galaxies, are in the range of thousands of kilometers per second.

It is not that we are the center of the universe, either. Observers in any galaxy would be seeing the same phenomenon, each galaxy moving apart like the raisins in bread that is rising. This expansion is a phenomenon at the largest scale. Locally, within our Local Group and local cluster, the galaxies are bound together by gravity and not engaged in any stampede.

Although the Big Bang did not take place in a localized spot like in the artist's impression shown here, it can be thought of as a truly titanic explosion that began the universe and set it in motion. NASA

What would an accelerating, expanding universe look like to an observer on Earth (or any other place)? The changes, of course, would be very, very gradual. The skies would darken first, though not to the naked eye, since we can only see a few galaxies without a telescope anyway. Eventually, though, the stars in the Milky Way would, in the normal course of things, reach the ends of their lives. Made increasingly of dead hulks, our galaxy itself would darken.

Ordinary Expansion?

In this scenario, now outmoded, stars would continue to be born, galaxies would continue to aggregate; but gradually, slowly, even this would end. Since dying stars do not release all their material back into interstellar space to form new stars—some of it remains locked up in their dead bodies—star formation would slow. As distant galaxies increased their distance from each other, even the dynamism of galactic collisions and mergers would become less and less frequent. This scenario was often called the Big Chill.

What would it look like? The sky would become darker and darker—very slowly, even more slowly than in an accelerating expansion, but with an occasional burst of fireworks as one of the increasingly rare star collisions or galactic collisions occurred somewhere nearby. (These collisions could prove quite disruptive, though, with some objects tossed out of their galaxies and some shoved inward, toward the black hole.) Eventually, complete darkness would descend.

Astronomers now believe that this scenario is not accurate. It has been replaced by the even more dramatic process of accelerating expansion described in the previous section.

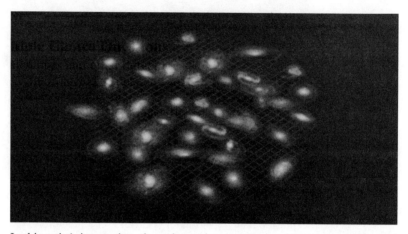

In this artist's impression, the universe is expanding and has formed stars and galaxies, but it is still strongly under the influence of the tug of gravity between all of the matter in galaxies and galaxy clusters. NASA

Instead of slowing down, the expansion of the universe is now know to have started to accelerate. This will cause the distances between galaxies to grow at an ever increasing rate in the future. NASA

Recollapse?

In this scenario—once nicknamed the Big Crunch—gravity would overwhelm dark energy (or dark energy would somehow turn out not to exist at all). The universe would be pulled inward, galaxies moving closer and closer together, imploding like the Big Bang's explosion, but in reverse. The result would be a black hole, nearing the end of spacetime.

This recollapse, one thought plausible, has now been discredited. It no longer matches theory or observations.

Close to Steady State?

For the universe to remain pretty much the way it is now, a comfortable "coasting universe" or one that expands but at a vastly slower and slower rate, an unlikely situation would have to exist: that we, in the early years of the twenty-first century, as measured on a small planet around a medium-sized star in an exurb of an ordinary spiral galaxy, happen to be living at a poised point in the evolution of the universe. One would not place a large bet that this is the case... unless of course, the universe never does change much, never has, and never will.

The universe did indeed begin at a poised point of sorts, not too dense to cause a collapse early on, but dense enough for galaxies to form. Astronomers know, however, that the universe has changed a great deal over the 13.7 billion years since then. Mass has become distributed unevenly because of the existence of galaxies, for example. So a steady state universe is highly unlikely—the future will almost certainly continue to surprise us.

Last Stages

The very last stages of what will probably be an accelerating, expanding universe will be populated mostly by black holes and protons (a subatomic particle).

Black holes will be at their most massive near the end since they will be the corpses of whole, dead galaxies (and even larger aggregations). Then even they will begin to evaporate, **slowly at first then** in massive bursts of radiation called Hawking radiation. What will be left will be particles and subatomic particles, sprayed all over the universe, no longer gathered together in atoms and molecules—or anything larger.

The protons themselves may have decayed even earlier than this. Astronomers are searching now for any evidence they can find of proton decay. It is possible that a low level of decay already exists.

The end result would be nothingness.

In this chapter you have learned that:

✔ Of the four scenarios for the denouement of the universe, accelerating expansion is now looking to be by far the most likely.

✔ Uncertainties about the mass of dark matter and the power of dark energy, however, remain.

✔ At the end of time, probably all that will be left is widely scattered, decayed, subatomic particles

True/False Questions

1. Astronomers now think that the most plausible future for the universe involves an accelerating expansion.

Completion Questions

1. At the very end the _____ will evaporate.

Multiple Choice Questions

1. The battle that will determine the nature of the universe's future is first between
   a. Dark energy and gravity
   b. The galaxies
   c. Protons and antiprotons

ANSWER KEY

True/False

1. T

Completion Questions:

1. Black holes

Multiple Choice Questions:

1. a

Frontiers in Astronomy

You should read this chapter if you need to review or learn about:

 Telescopes and other detectors astronomers are using to study the universe at different wavelengths

 How these devices are leading, and will lead in the future, to more knowledge about our solar system, stars and galaxies and their evolution, exoplanets in particular, dark matter and dark energy, and more

 How space voyages can also extend our understanding

 Some of the most important astronomical areas we know far too little about

Get Started

Human curiosity about space has led to the invention of telescopes of many kinds, from Galileo's early spyglass to massive systems of mirrors moved into space by rockets. Though some are based on the Earth's surface, quite a few are aloft in near-Earth orbit, with more of these in the works. Some of these detectors open a broad eye to space phenomena, while others specialize in examining astronomical objects at a specific wavelength. The information this book is based upon has been gleaned by astronomers in all of these ways.

An Array of Detectors

Probably the most famous major telescope today is the *Hubble Space Telescope*. Aloft for more than 15 years above the thick blanket of the Earth's atmosphere, it has clearly imaged marvels from the births of distant stars to the spectra, or light profile, of supernovas, to the newest exoplanets. Its specialties have been visible light and ultraviolet light, as well as gathering the most precise spectra possible.

The Hubble's base out in space, in near-Earth orbit, has been the key to its success. The Earth's atmosphere blocks out almost all forms of radiation except for visible light (which is fortunate for us since some of the wavelengths of radiation are lethal to humans). And, as this telescope reaches the end of its useful lifetime, some of the other space telescopes already in space and collecting information will become even more important. Four of the major ones are:

- *Chandra X-ray Observatory*. Aloft since 1999 and to be in operation for at least several more years, Chandra has discovered the supermassive black holes of distant galaxies by their x-ray radiation, suggested the existence of medium-sized black holes, shown the dynamic emissions and magnetic fields of neutron stars, analyzed the effluents of supernovas, mapped the massive areas of hot gas among galactic clusters, imaged the pulsating jets of pulsars, and more. Its specialty is in the x-ray wavelength.

- *GALEX, the Galaxy Evolution Explorer*. Launched into space in 2003 and specializing in the ultraviolet light portion of light's spectrum, GALEX has a significant lifetime ahead of it. So far, it has mapped star formation in galaxies near to ours. Farther out—over the last 10 billion years of spacetime—it has established how far into galactic disks the newest, most massive, short-lived stars are born. It has also used the lack of ultraviolet light emitted by ancient quasars to see whether star formation slows down considerably as galaxies grow older.

- *Spitzer Space Telescope*. Looking in the (ultraviolet) **infrared** portion of the spectrum, the Spitzer stares at the youngest objects, seeing right through the dust clouds that often surround them. So far, it has found large amounts of organic compounds (the building blocks of life from water to methane to carbon dioxide) in a galaxy more than 3 billion light-years away. Far nearer—at only 25 light-years distant—it is zeroing in on a potential debris disk or exoplanet zone. Plenty of mapping of the Milky Way's dust-shrouded inner bulge, and how dust concentrations change in general, are part of its workload. And the Spitzer seems to be establishing the upper limit for the weight of any star—about 150 solar masses. This telescope should last until about 2009.

- *Swift Gamma Ray Burst Explorer*. Launched into space in late 2004, the Swift has been looking at how the powerful and not-well-understood gamma ray bursts of power originate, differ among themselves, and signal conditions in the early universe. These bursts, described in Chapter 13, The Death of Stars, produce as much energy in a second or so as the Sun would, in total, if it lived for 80 billion years.

Where one looks within the spectrum of light—gamma ray/x-ray/ultraviolet/visible light/infrared/radio—determines what one sees. A galaxy seen in its visible radiation can look completely different when analyzed in a different wavelength, since the objects emitting significantly in that wavelength stand out. Visible light can show main sequence stars and the gas and dust clouds illuminated by them, for example, but in a gamma ray snapshot the powerful gamma ray bursts can become evident (they are described in Chapter 13, The Death of Stars). The x-ray wavelength can point out the hot gas clouds (between) **within** galactic clusters which are key to the nature of dark energy. Ultraviolet light devices show us the newest, hottest stars. And infrared radiation illuminates the galaxies farthest away from us in spacetime. These are only a few examples of the different slices of the universe that the different kinds of detectors can reveal.

Knowledge at the Frontier

The telescopes described above, and many more detectors, are pushing at the frontiers of our knowledge about our solar system, star and galactic evolution, exoplanets, and dark energy/dark matter. A few selected examples of these investigations—both orbiting telescopes and space probes—now and planned in operations for the future, are described next.

Solar System Probes

- *Ulysses*. Aloft since 1990, this probe's orbit swings from high above the poles of the Sun to near Jupiter and back, and it has detected significant amounts of (dust from other stars) **particles from interstellar space** entering our own solar system. It is also analyzing how the Sun's magnetic field flips its polarity, a phenomenon responsible for the Sun's cycle of sunspots that increases and decreases the occurrence of aurora borealis and aurora australis on Earth.

- *SOHO, the Solar and Heliospheric Observatory*. This is our space weather observatory, anchored between the Earth and the Sun, a spacecraft which has been watching our active star since the mid-1990s. It provides a 1–3 day notice (depending upon the solar phenomenon) of the changing nature of the solar wind approaching the Earth. Large blasts would harm any astronauts who happened to be in space at the time, and they regularly disrupt Earth's communication satellites. Those interested in a forecast of the Northern Lights or Southern Lights can go to www.spaceweather.com to find one.

> Pioneer 10, in memoriam. The first spacecraft ever to reach the edge of our solar system, it sent its last signal to us in 2003. Launched back in 1972, it swept out and through the asteroid belt, near Jupiter, out past Pluto, and to the edge of the "heliopause," the end of the influence of the Sun via its solar wind, taking pictures all the way. It is now inactive, somewhere in interstellar space—never to return.

- Rosetta. Launched in 2004, this spacecraft will reach, and land upon, a selected comet in our solar system, even burrowing down toward the nucleus. The landing date is to be late 2014.

Star and Galactic Evolution Frontiers

- Hubble Space Telescope. Among the many upcoming explorations and findings of the Hubble in these areas of astronomy are its growing atlas of the distribution of galaxies in the universe and the spectra of possible exoplanets as they pass in front of their stars from our vantage point on Earth. It is surveying thousands more stars in search of more exoplanets, watching the shock waves of a nearby supernova, and scanning more distant supernovas to establish more about the nature of dark energy.

- 2 MASS. Positioned on Earth, one part of it in the Northern and one in the Southern hemisphere, and observing in the infrared wavelengths, this pair of telescopes has now placed online an atlas of 5 million images from its celestial mapping activities, called the Two Micron All-Sky Survey (2 MASS). Look for it at www.ipac.caltech.edu/2mass/gallery.

- Two important new telescopes will launch and become operational in about 2010–2013. One is, unusually, the project of a television channel: the Discovery Channel Telescope, a detector launched in partnership with the Lowell Observatory, is designed to bring in images from the outer solar system and beyond. The second, a major scientific undertaking, is the James Webb Space Telescope. It will look in the infrared wavelengths for the earliest, oldest stars and galaxies, for example, and will attempt to fill the shoes of the Hubble.

- A bit later than the above, a new array of radio telescopes should be online. These are to be huge Earth-surface "dishes," situated in various places around the globe; they are designed to collect information from objects emitting strongly in the radio part of the spectrum. (This has nothing to do with radio stations, though these transmit using radio waves also.) Online already in this domain is the VLA (Very Large Array), which has brought us information about strange bursts from near our galactic core. The new, larger array should provide even more information along these lines.

- A decade or so in the future, more new telescopes will be constructed to investigate supermassive black holes, the history of their formation, the first observational gravity-wave studies, and galactic distances. These are NASA's Constellation-X, the European Space Agency's X-ray Evolving Universe Spectrometer, the Laser Interferometer Space Antenna (LISA), and the European Space Agency's Gaia spacecraft.

The best way to find the very latest on launch schedules, by NASA, is to go to http://www.nasa.gov/missions/highlights/schedule.html.

Exoplanets

- Just up and running is the Earth surface-based Large Binocular Telescope (LBT). Among other projects, LBT will be looking for exoplanets.

- Scheduled to launch in 2008 is NASA's Kepler spacecraft. It will search intensively for exoplanets of all sizes and at all distances from their stars. See the relevant sections in Chapter 5, The Solar Nebula and the Formation of Gas Planets, for background on exoplanets.)

- By about 2016, the largest telescope on Earth, or space, is to be at work. Called the Giant Magellan Telescope, its lens will be 83.3 feet (25.4 meters) in diameter, allowing it to collect faint light from everything from exoplanets to the rings around black holes to the most distant (oldest) galaxies of all.

Dark Matter/Dark Energy

- Large Hadron Collider (LHC). In 2007, this new particle accelerator should be online and looking for a new type of neutrino, key to furthering the understanding of dark energy.

- The search for neutrinos is already ongoing at several Earth-surface sites. Among these is the Antarctic detector Arctic Muon and Neutrino Detector Array II (AMANDA II). (Its first version was sited in the Arctic.) Since neutrinos from the Sun and even from far-flung black hole collisions pass through all matter, they can be detected as they sweep through masses of deep buried ice. Nothing else can enter there to interfere with their signals.

- Gamma-ray Large Area Space Telescope (GLAST), a space-based telescope, is also to be up in late 2007. It will search for many things within the gamma ray portion of the spectrum, including faint emissions from dark matter in the halo of our galaxy.

- Supernova Acceleration Probe (SNAP). Set for a 2010 launch but with its funding currently in doubt, this probe is to take the spectra of some 6000 of the Type 1a supernovas whose distance is crucial to analyzing the behavior, over time, of dark energy. (As mentioned in Chapter 16, The Role of Dark Energy, these objects are absolutely central to this analysis.)

- Dark Universe Observatory. This set of seven telescopes will look for x-ray patterns to indicate how long ago superclusters of galaxies were able to form. This level of clustering is crucial to the understanding of dark energy.

Space Voyages

Though designing, constructing, and positioning a telescope or other detector can be extraordinarily expensive, an actual space voyage, robotic or with a human component, is usually much more so. Though there are necessarily fewer of them, voyages have led to, and will lead to, an extension of our understanding of space phenomena. Future voyages, in near space, to the Moon, to Mars, and beyond, are sampled next, along with the issues associated with them.

- Space Shuttles. In the works for these near-Earth orbiters of the future are a small emergency escape vehicle which could also be used as a ferry for astronauts, as well as a new style of shuttle that uses fuels such as solid paraffin wax and oxygen scooped up from our atmosphere on its way up. (Both the weight and the combustibility of the fuels currently used add extravagantly to the expense and somewhat to the danger of space flight.)

- Back to the Moon? After 30-some years, it may be time to go back, and the funding may be there to do so. The moon—outfitted with a Moon base—could be an excellent place to test technologies such as a fueling station (the fuel derived from moon rocks and soils) and human shelters (complete with melted Moon ice for water and solar energy) that could also be used as a base for explorations to points further away. Studying the Moon from a moon-orbiting observatory is also a desirable goal because it is so similar to our Earth early in its history.

- On to Mars? The robotic probes that crawled over parts of the Martian planet at the middle of this decade whetted the appetites of many scientists and curious citizens. Yes, it appears that Mars once had water. That means that there are places which could be searched for microbial life as we know it, or at least for its fossils, on a next voyage robotic or human. Fuels under investigation include higher efficiency nuclear propulsion, which could also take us beyond Mars, and solar power (current solar cells do not perform well at greater distances from the Sun). Other research will focus on protecting any astronauts from vigorous blasts of the solar wind; on long voyages these would occur occasionally and could even be life-threatening.

- Beyond? For more distant travel, astronomers are learning how to use the gravity of one planet as a slingshot to another and have even discovered natural "superhighways" within the solar system where gravitational pulls and pushes can lead to amazing fuel efficiencies. Already this method is being used in planning the route for the Jupiter Icy Moons Orbiter (JIMO) due to launch in 2012. It will explore four of the Jovian planet's moons for water and other properties. Use of solar sails—thin sails fitted to use faint amounts of solar energy and the solar wind—are under investigation too, along with much more.

- Well Beyond? Will there be voyages to the stars and their exoplanets? Will you volunteer to go? To help develop the probes and plan the voyages?

Unknowns

The magnetic lure of what is not known pulls science forward. Generally, the more one does know, the more unknowns keep emerging.

Throughout the book, a few of the key areas in which we need to know more now have been mentioned. A brief list:

- How does gravity work with electromagnetism and the other two (quantum mechanical) forces of nature, the weak force and the strong force? The search is on for gravity waves and for the integrative mathematics here.
- Why do galaxies form where they do?
- How long has galactic superclustering been going on?

- Is the Milky Way a barred-spiral galaxy and, if so, what is the size of our central bar?

- What processes are going on at the event horizons of black holes and at their central singularities?

- In black holes, pulsars, and other astronomical objects, how are the forceful jets of radiation emitted and directed?

- How much material—ordinary, visible matter as well as dark matter—lies between superclusters of galaxies?

- What is dark matter made of?

- Are there additional subatomic particles beyond those which we have detected?

- How strong is dark energy? Does it truly exist?

- What is the future of the universe as we know it?

In this chapter you have learned that:

 Earth surface–based telescopes and other detectors (including particle detectors), along with those aloft in near-Earth orbit, are what drive observational knowledge forward in astronomy.

 Many more such detectors and probes are in the works.

 Space voyages, robotic or human, will also continue, though there are fewer of these since they are even more expensive.

 The unknowns in astronomy are at least as significant as what is now known.

Test Yourself

True/False Questions

1. Telescopes and other detectors are placed in near-Earth orbit to see objects and phenomena blocked by our atmosphere.

Completion Questions

1. With SOHO, a space weather observer, it is possible to predict _____.

Multiple Choice Questions

1. Telescopes and other detectors
   a. Are protected against all forms of radiation
   b. Often specialize in one wavelength of light
   c. Include the active Pioneer 10 probe

ANSWER KEY

True/False
1. T

Completion Questions
1. The Northern Lights and Southern Lights

Multiple Choice Questions
1. b

Index